Conservation and Reintroduction of Rare and Endangered Plants in China

Hai Ren
Editor

Conservation and Reintroduction of Rare and Endangered Plants in China

Springer

2020 UN BIODIVERSITY CONFERENCE
COP 15 · CP/MOP10 · NP/MOP4
Ecological Civilization-Building a Shared Future for All Life on Earth
KUNMING · CHINA

Editor
Hai Ren
South China Botanical Garden, CAS
Guangzhou, China

ISBN 978-981-15-5300-4 ISBN 978-981-15-5301-1 (eBook)
https://doi.org/10.1007/978-981-15-5301-1

This Springer imprint is published by the registered company Springer Nature Singapore Pte Ltd.
The registered company address is: 152 Beach Road, #21-01/04 Gateway East, Singapore 189721, Singapore

Preface

In the face of unprecedented biodiversity losses and global change, effective approaches for the conservation of rare and endangered plant species are urgently required. The major approaches to integrative plant conservation include in situ conservation, ex situ conservation, and reintroduction. Reintroduction may be especially effective at protecting and rescuing rare and endangered plants.

China has protected about 65% of the vascular plant communities through in situ conservation in natural reserves and national park systems and has preserved about 60% of the plant species through ex situ conservation in botanical gardens and other ex situ conservation facilities. However, we know less about reintroduction in China. Throughout the book, we and our invited authors explore to what extent information about reintroduction of plants is currently available in China.

This book is composed of two parts. Part I introduces the plant diversity and its conservation in China, and Part II displays some cases of reintroduction of rare and endangered plants in China. The majority of the chapters in the book are devoted to the case studies of reintroduction.

Books such as this one become a reality only with the support and involvement of many people. The editors are indebted to the team of enthusiastic authors, all famous experts in the fields in China, who have made available experience. The editors divide their work as follows: Prof. Hai Ren designed the contents and organized the manuscripts. Prof. Hongfang Lu edited Chaps. 1–4, Dr. Hongxiao Liu edited Chaps. 5–20, Director Ju Zhou co-organized the manuscripts and co-conceived the contents, Prof. Yan Zeng edited tables, photos, and figures. Thanks to Prof. Elizabeth Platt Hamblin for editing English. We thank the anonymous reviewers for their constructive comments. We are very grateful to Dr. Xin Zhu and Beracah John Martyn for their careful fine-tuning of the editorial work at press.

The financial support from Chinese Academy of Sciences and Ministry of ecological environment (No. 8-3-7-20-10) is gratefully acknowledged. We hope that this small book will be of value in stemming the tide of plant diversity loss and unsustainable development in China and even in the world. This book is dedicated to Convention on Biological Diversity-COP 15, which will be held in 2021 in Kunming, China.

Guangzhou, China
November 12, 2019

Hai Ren

Acknowledgements

Prof. Hongfang Lu, Dr. Hongxiao Liu, Director Ju Zhou, and Prof. Yan Zeng are Associate Editor-in-Chief.

Prof. Hongfang Lu edited Chaps. 1–4, Dr. Hongxiao Liu edited Chaps. 5–20, Director Ju Zhou co-organized the manuscripts and co-conceived the contents, Prof. Yan Zeng edited tables, photos, and figures.

Contents

Part I Plant Diversity and Its Conservation in China

An Overview of China's Vegetation and Plant Diversity 3
Haining Qin, Xiaohua Jin, and Ke Guo

Rare and Endangered Plants in China 21
Haining Qin, Xiaohua Jin, and Lina Zhao

In Situ and Ex Situ Conservation of Chinese Plants 33
Xiaohua Jin and Hai Ren

Reintroduction of Rare and Endangered Plants in China 49
Hai Ren, Hongfang Lu, Hongxiao Liu, and Zhanhui Xu

Part II Reintroduction Cases of Rare and Endangered Plants in China

Reintroduction of *Primulina tabacum* Hance, a Critically Endangered Calciphilous Perennial Herb, in Southern China 111
Hai Ren, Guohua Ma, Qianmei Zhang, and Xiangying Wen

Artificial Propagation and Reintroduction of *Dayaoshania cotinifolia* W. T. Wang, an Extremely Small Population in China 117
Guohua Ma, Beiyi Guo, Songjun Zeng, Yubing Wang, Dehua Tu, Jinfeng Lü, and Hai Ren

Reintroduction of *Tigridiopalma magnifica* C. Chen, a Critically Endangered Endemic Plant in China 125
Songjun Zeng, Hai Ren, Longna Li, and Kunlin Wu

Reintroduction of *Camellia changii* Ye (Theaceae), a Critically Endangered Plant Endemic to Southern China 133
Hai Ren, Hongxiao Liu, Qianmei Zhang, Yi Xu, Ju Zhou, and Yan Zeng

Conservation Translocation of the Rare and Endangered Tree *Euryodendron excelsum* in South China 139
Hai Ren, Qianmei Zhang, Yiming Fan, Zeyuan Zou, and Yi Xu

Conservation Translocations of *Manglietia longipedunculata*, a Critically Endangered Tree in South China 145
Hai Ren, Hongxiao Liu, Qianmei Zhang, Yan Zeng, Ju Zhou, Wenhui Zhu, and Xiaodong Cui

The Reintroduction of *Parakmeria omeiensis* Cheng, a Critically Endangered Endemic Plant, in Southwest China 151
Daoping Yu, Xiangying Wen, and Cehong Li

Reintroduction of *Hopea hainanensis* Merr. and Chun, a Wild Plant with Extremely Small Populations, in Southern China 159
Runguo Zang, Yi Ding, Jihong Huang, and Yue Xu

Reintroduction of *Trigonobalanus doichangensis*, a Threatened Plant Species with Extremely Small Populations 167
Weibang Sun, Fengrong Li, and Yuan Zhou

Reintroduction of *Craigia yunnanensis* to Private Plots: An Innovative Case with Help from Local Volunteers 175
Jing Yang, Yaling Chen, and Weibang Sun

Reintroduction of *Myricaria laxiflora* (Franch.) P.Y. Zhang et Y.J. Zhang, a Critically Endangered Shrub, in Central China 181
Zhiqiang Xiao and Mingxi Jiang

Reintroduction of *Bretschneidera sinensis* Hemsl., an Endangered Endemic Tree 187
Hongfeng Chen, Qi Qiao, Hailin Liu, Xiangying Wen, and A. J. Harris

In Vitro Propagation and Reintroduction of *Paphiopedilum wardii* Sumerh. 193
Songjun Zeng, Kunlin Wu, Zhilin Chen, and Jun Duan

Reintroduction and Assisted Colonization of *Paphiopedilum spicerianum*: An Orchid with Extremely Small Populations in China . . . 199
Jiangyun Gao, Xuli Fan, and Wenke Yang

Orchid Conservation Translocation Efforts in China 205
Hong Liu and Jiangyun Gao

Reintroduction of Wild *Cycas* Species in China 225
Shuguang Jian, Xun Gong, and Nian Liu

Contributors

Hongfeng Chen South China Botanical Garden, Chinese Academy of Sciences, Guangzhou, China

Yaling Chen University of Chinese Academy of Sciences, Chinese Academy of Sciences, Beijing, China
Chinese Academy of Sciences, Kunming Institute of Botany, Kunming, Yunnan, China

Yongju Chen Administration Ehuangzhang Nature Reserve, Yangchun, Guangdong, China

Zhilin Chen South China Botanical Garden, Chinese Academy of Sciences, Guangzhou, China
Horticulture Research Institute, Guizhou Academy of Agricultural Sciences, Guiyang, China

Xiaodong Cui Administration of Nankunshan Nature Reserve, Longmen, Guangdong, China

Yi Ding Research Institute of Forest Ecology, Environment, and Protection, Chinese Academy of Forestry, Beijing, China

Jun Duan South China Botanical Garden, Chinese Academy of Sciences, Guangzhou, China

Xuli Fan Yunnan Key Laboratory of Plant Reproductive Adaption and Evolutionary Ecology, Yunnan University, Kunming, Yunnan, China
Laboratory of Ecology and Evolutionary Biology, Yunnan University, Kunming, Yunnan, China

Yiming Fan Foshan Institute of Science and Technology, Foshan, China

Jiangyun Gao Yunnan Key Laboratory of Plant Reproductive Adaption and Evolutionary Ecology, Yunnan University, Kunming, Yunnan, China
Laboratory of Ecology and Evolutionary Biology, Yunnan University, Kunming, Yunnan, China

Xun Gong Chinese Academy of Sciences, Kunming Institute of Botany, Kunming, Yunnan, China

Beiyi Guo South China Botanical Garden, Chinese Academy of Sciences, Guangzhou, China
University of Chinese Academy of Sciences, Chinese Academy of Sciences, Beijing, China

Ke Guo Institute of Botany, Chinese Academy of Sciences, Beijing, China

A. J. Harris South China Botanical Garden, Chinese Academy of Sciences, Guangzhou, China

Jihong Huang Research Institute of Forest Ecology, Environment, and Protection, Chinese Academy of Forestry, Beijing, China

Mingxi Jiang Chinese Academy of Sciences, Wuhan Botanical Garden, Wuhan, China

Shuguang Jian South China Botanical Garden, Chinese Academy of Sciences, Guangzhou, China

Xiaohua Jin State Key Laboratory of Systematic and Evolutionary Botany, Institute of Botany, Chinese Academy of Sciences, Beijing, China

Cehong Li Sichuan Provincial Institute of Natural Resource Science, Chengdu, Sichuan, China
Emeishan Botanical Garden, Emeishan, Sichuan, China

Fengrong Li Chinese Academy of Sciences, Kunming Institute of Botany, Kunming, Yunnan, China

Longna Li South China Botanical Garden, Chinese Academy of Sciences, Guangzhou, China
Laboratory Center of Life Sciences, Nanjing Agricultural University, Nanjing, China

Hailin Liu South China Botanical Garden, Chinese Academy of Sciences, Guangzhou, China

Hong Liu Center for Tropical Plant Conservation and Department of Earth and Environment, Florida International University, Miami, FL, USA
Fairchild Tropical Botanic Garden, Coral Gables, FL, USA

Hongxiao Liu South China Botanical Garden, Chinese Academy of Sciences, Guangzhou, China

Nian Liu Zhongkai University of Agriculture and Engineering, Guangzhou, China

Hongfang Lu South China Botanical Garden, Chinese Academy of Sciences, Guangzhou, China

Jinfeng Lü South China Botanical Garden, Chinese Academy of Sciences, Guangzhou, China

Guohua Ma South China Botanical Garden, Chinese Academy of Sciences, Guangzhou, China

Qi Qiao South China Botanical Garden, Chinese Academy of Sciences, Guangzhou, China

Haining Qin State Key Laboratory of Systematic and Evolutionary Botany, Institute of Botany, Chinese Academy of Sciences, Beijing, China
College of Life Sciences, University of Chinese Academy of Sciences, Beijing, China

Hai Ren South China Botanical Garden, Chinese Academy of Sciences, Guangzhou, China

Weibang Sun Chinese Academy of Sciences, Kunming Institute of Botany, Kunming, Yunnan, China

Dehua Tu Laoshan Nature Reserve of Jinxiu in Guangxi Zhuang Nationality Autonomous Region, Jinxiu, Guangxi, China

Yubing Wang China Three Gorges University, Yichang, Hubei, China

Xiangying Wen South China Botanical Garden, Chinese Academy of Sciences, Guangzhou, China
Botanical Garden, Chinese Academy of Sciences, Guangzhou, China

Kunlin Wu South China Botanical Garden, Chinese Academy of Sciences, Guangzhou, China

Zhiqiang Xiao Chinese Academy of Sciences, Wuhan Botanical Garden, Wuhan, China

Yi Xu Administration Ehuangzhang Nature Reserve, Yangchun, Guangdong, China

Yue Xu Research Institute of Forest Ecology, Environment, and Protection, Chinese Academy of Forestry, Beijing, China

Zhanhui Xu South China Botanical Garden, Chinese Academy of Sciences, Guangzhou, China

Jing Yang Chinese Academy of Sciences (CAS), Kunming Institute of Botany, Kunming, Yunnan, China

Wenke Yang Yunnan Key Laboratory of Plant Reproductive Adaption and Evolutionary Ecology, Yunnan University, Kunming, Yunnan, China
Laboratory of Ecology and Evolutionary Biology, Yunnan University, Kunming, Yunnan, China

Daoping Yu Sichuan Provincial Institute of Natural Resource Science, Chengdu, Sichuan, China
Emeishan Botanical Garden, Emeishan, Sichuan, China

Runguo Zang Research Institute of Forest Ecology, Environment, and Protection, Chinese Academy of Forestry, Beijing, China

Songjun Zeng South China Botanical Garden, Chinese Academy of Sciences, Guangzhou, China

Yan Zeng Bureau of Science and Technology for Development, Chinese Academy of Sciences, Beijing, China

Qianmei Zhang South China Botanical Garden, Chinese Academy of Sciences, Guangzhou, China

Lina Zhao State Key Laboratory of Systematic and Evolutionary Botany, Institute of Botany, Chinese Academy of Sciences, Beijing, China
College of Life Sciences, University of Chinese Academy of Sciences, Beijing, China

Ju Zhou Bureau of Science and Technology for Development, Chinese Academy of Sciences, Beijing, China

Yuan Zhou Chinese Academy of Sciences, Kunming Institute of Botany, Kunming, Yunnan, China

Wenhui Zhu Administration of Nankunshan Nature Reserve, Longmen, Guangdong, China

Zeyuan Zou South China Agricultural University, Guangzhou, China

Part I
Plant Diversity and Its Conservation in China

An Overview of China's Vegetation and Plant Diversity

Haining Qin, Xiaohua Jin, and Ke Guo

Abstract China's complex climate and geological condition have enriched species and created a diversified spatial distribution pattern of biodiversity, making China one of the countries of the most abundant of species and special types of vegetation in the world. Seven vegetation type groups are recorded by *Vegetation of China* (Editorial Board of Vegetation of China, *Vegetation of China* (Science Press, Beijing, 1980)): forests, shrubs, deserts, grasslands, meadows, alpine vegetation, and swamp or aquatic vegetation. The main characteristics of Chinese vegetation are (1) a complete range of categories, including various types of forests, steppes, deserts, and marshes; (2) vast areas of subtropical evergreen broadleaf forest owing to the influence of the monsoon climate; and (3) unique plateau vegetation on the Qingzang Plateau and the unique altitudinal spectra of numerous mountains. The complexity of China's vegetation types is not only a manifestation of China's complex biodiversity but also is the basis for the evolution of diverse plant species, animals, fungi, and microorganisms. China has 36,152 species of higher plants, 9% of the world flora. Two national floras completed in early of this century *Flora Reipublicae Popularis Sinicae* (1959–2004) and *Flora of China* (1994–2013) are the key breakthroughs in understanding and mastering China plant diversity. *Catalogue of Life China* (CoL-China) (2008–) is designed to update the species information in the above floras with latest taxonomic literatures, to meet timely the needs of biodiversity conservation. These three works are recognized as the primary reference

H. Qin (✉)
State Key Laboratory of Systematic and Evolutionary Botany, Institute of Botany, Chinese Academy of Sciences, Beijing, China

College of Life Sciences, University of Chinese Academy of Sciences, Beijing, China
e-mail: hainingqin@ibcas.ac.cn

X. Jin
State Key Laboratory of Systematic and Evolutionary Botany, Institute of Botany, Chinese Academy of Sciences, Beijing, China
e-mail: xiaohuajin@ibcas.ac.cn

K. Guo
Institute of Botany, Chinese Academy of Sciences, Beijing, China

H. Ren (ed.), *Conservation and Reintroduction of Rare and Endangered Plants in China*, https://doi.org/10.1007/978-981-15-5301-1_1

book for the study of China plant diversity. Hundreds of new species are described every year and added to the annual checklist of CoL-China. It is essential to complete inventory of Chinese plants, since it's impossible to conserve the undiscovered species.

Keywords Mega-diversity country · Vegetation · Plant species diversity · Conservation

1 An Overview of China's Vegetation

From south to north, the eastern monsoon region of China has tropical forests and seasonal rain forests; subtropical evergreen broadleaf forests; warm temperate deciduous broadleaf forests; temperate coniferous, broadleaf mixed forests; and cold temperate coniferous forests, showing obvious latitudinal gradients. From east to west, north China gradually transitions from forest to grassland and then to desert along longitudinal gradients (Fig. 1).

The vegetation on the Qinghai-Tibet Plateau is a variation on highland integrating the horizontal and vertical gradients, with a distribution pattern as follows: from the southeast to the northwest, with the gradual increase of altitude, there is a sequential distribution of a mountain forest belt (Fig. 2), a belt of alpine shrub land and meadows, an alpine grassland belt, and an alpine desert belt.

Fig. 1 Some vegetation types in China

Fig. 2 Some mountain forests in China

1.1 Forests

Forest is a type of vegetation in which trees dominate and the canopy coverage reaches 30% or more. Its community structure is usually complex; for example, the canopy can be divided into two or more sub-layers, under which there are usually shrub layers, then herb, and moss layers. The forest is mainly distributed in the eastern monsoon humid region and the mountains in semiarid and arid regions.

China's forest ecosystems are distributed from south to north in different types. Different climatic zones have their own representative forest types, including deciduous coniferous forests, evergreen coniferous forests, mixed coniferous and broadleaf forests, deciduous broadleaf forests, evergreen deciduous broadleaf mixed forests, evergreen broadleaf forests, rainforest and monsoon forests, and so on.

Deciduous coniferous forest is a forest type dominated by deciduous conifers, such as *Metasequoia glyptostroboides*, *Glyptostrobus pensilis*, *Pseudolarix kaempferi*, and *Taxodium distichum*. The main feature of this forest type is that the canopy layer defoliates collectively, forming a distinct growing season and a falling season within 1 year. According to the climatic characteristics of the distribution sites, China's deciduous coniferous forest is divided into two vegetation subtypes, namely, the subtype of cold-temperate deciduous coniferous forests featured by dominant *Larix* species and the subtype of warm-temperate deciduous coniferous forests featured by some species of Pinaceae and Taxodiaceae.

Evergreen coniferous forest is characterized by evergreen conifers as dominant species, in which the main feature is a green or gray-green appearance throughout the year. This type has a wide range of distribution and great variety of types of dominant species and communities. According to the climatic characteristics of community distribution, it is divided into four subtypes, namely, a cold-temperate evergreen coniferous forest subtype, a temperate evergreen coniferous forest subtype, a warm-temperate evergreen coniferous forest subtype, and a tropical evergreen coniferous forest subtype.

Mixed coniferous and broadleaf forest is composed of evergreen conifers and broadleaf trees. Usually, the conifers occupy the upper layer of the arbor layer, and the lower layer consists of more broadleaf trees. According to community characteristics such as community composition and the climate characteristics of the distribution area, it is divided into two subtypes of vegetation, namely, the typical coniferous and deciduous broadleaf mixed forest subtype distributed in the middle temperate zone and the mountainous conifer and broadleaf mixed forest subtype in the subtropics.

Deciduous broadleaf forest is dominated by deciduous broadleaf arboreal species. It is mainly distributed as zonal vegetation in low mountain and hilly areas in the mid-temperate and warm-temperate zone, and also as vertical zonal vegetation in temperate and subtropical mountains. There are many species that make up the arboreal layer, such as *Quercus* spp. (Fagaceae). Due to defoliation in winter, the seasons of the community are very distinct, with sufficient light under the canopy in winter and spring, allowing development of shrub and herb layers. According to the species composition characteristics of the community, the geographical distribution of the dominant species, and the characteristics of the habitat, it is divided into four vegetation subtypes, namely, the cold-temperate deciduous broadleaf forest subtype, the temperate deciduous broadleaf forest subtype, the warm-temperate deciduous broadleaf forest subtype, and desert riverbed deciduous broadleaf forest subtype.

Evergreen deciduous broadleaf mixed forest is a transitional forest type between deciduous broadleaf forests and evergreen broadleaf forests. In China, it is widely distributed in the northern subtropical regions and subtropical limestone mountains. The composition of the community is diverse. From north to south, the evergreen components are gradually increasing, and the species also have substitution phenomena or more hot and evergreen components. There is usually no obvious dominant species, and the species are mainly in the families of Fagaceae, Betulaceae, and Ulmaceae. The community structure is complicated, and the deciduous tree species usually occupy the upper part of the arboreal layer, which makes the seasonal phase change obviously, while the evergreen tree species have a comparative advantage in the lower part of the arboreal layer. According to the difference of habitats, the evergreen deciduous broadleaf mixed forest is divided into three vegetation subtypes: the (zonal) evergreen deciduous broadleaf mixed forest distributed in the low hills of the northern subtropical zone, the mountain evergreen deciduous mixed broadleaf forest distributed in the subtropical mid-mountain belt, and the limestone evergreen deciduous broadleaf mixed forest distributed on calcareous soil.

Evergreen broadleaf forest is a vegetation type with evergreen broadleaf tree species as dominant species, widely distributed in subtropical regions with warm and humid climates and abundant precipitation and mountains in the north of the tropics. The species is diverse in the community, whose composition prone to be more complex from north to south. The dominant species are usually not very obvious, mainly including plants from Fagaceae, Lauraceae, and Theaceae families. The arboreal layer can often be further divided into different sub-layers. Plants in the shrub layer mainly include Ericaceae, Symplocaceae, Myrsinaceae, Melastomataceae, and Aquifoliaceae species, being usually sparse, with the height generally below 2 m; the layer also contains a large number of tree seedlings. The herb layer is mainly composed of Liliaceae, Cyperaceae, Zingiberaceae, and Chloranthaceae plants and ferns, distributed sporadically, all hygrophilous. The sporadic patches of moss are distributed on bare rocks, roots, and on base of trunks. Lianas and parasitic plants are also common. The evergreen broadleaf forest has a wide distribution area and various types, so the type division is complicated.

Tropical forests in China mainly include *tropical monsoon forests and tropical rain forests*. The former is a type of tropical forest distributed on the edge of the tropics, found in areas with alternating periodic wet and dry conditions. The latter is distributed in areas with abundant rainfall and sufficient heat, but it is located further north compared with the global distribution of rainforests. China's tropical forests are small in area, mainly distributed in southern Guangdong, Hainan Island, south-western Guangxi, southern Yunnan, and southeastern Tibet. They are the most abundant forest ecosystems in China.

1.2 Shrubs

Shrub assemblages are dominated by mesophytic and meso-xerophytic shrub species, with height generally lower than 5 m and coverage more than 30%. The community structure is relatively simple, usually containing shrub layers and herb layers and occasionally a few bryophytes under thick bushes. Natural shrubs are mainly distributed in humid and semi-humid environments where natural conditions, including temperature and moisture, are no longer suitable for the growth of arboreal species, such as forest-to-grass transition areas, wet areas above forest lines, and mountains with relatively sufficient water in arid areas, or secondary vegetation types in places where forest has been destroyed. China's shrubs can be divided into five vegetation types, the main and widely distributed types including deciduous broadleaf shrubs and evergreen broadleaf shrubs.

Deciduous broadleaf shrub is a vegetation type with deciduous shrubs as dominant species in the communities. Deciduous broadleaf shrub has a vast distribution, quite different habitats, many dominant species, and complicated community types. It has many primitives, while many of them being secondary vegetation formed after forest destruction. The height of the community is mostly between 1 and 2 m, some types reaching 4 m and some less than 1 m. The coverage is generally between 30 and 70%. According to the characteristics of community and habitat, it can be divided into multiple vegetation subtypes.

Evergreen broadleaf shrub is a vegetation type of evergreen broadleaf shrubs as dominant species. It is mainly secondary vegetation formed after the destruction of forests in tropical and subtropical regions, which is a relatively stable stage in the succession process. The community is generally composed of evergreen young trees and shrubs and sometimes mixed with some deciduous species. The shrub layer of most communities is 1–2 m high, and the coverage is above 60%. The herb layer is generally well developed, with a height of less than 1 m, while the coverage is greatly affected by the shrub layer. According to the ecological characteristics and habitat conditions of dominant species, it is also divided into various subtypes.

1.3 Deserts

Desert is a sparse ecosystem type developed in extreme drought habitats with rare precipitation and strong evaporation. It is a sparse vegetation composed mainly of extremely xeric shrubs, semi-shrubs, small semi-shrubs, and small semi-arboreal and succulent plants. China's deserts are concentrated in the inland areas of the northwest and belong to temperate deserts, accounting for about one-fifth of China's land area, including about one million km^2 of desert and Gobi. The dominant species are mainly from families such as Chenopodiaceae, Zygophyllaceae, Compositae, Caryophyllaceae, Ephedraceae, and Fabaceae. There are many types of communities, and the species composition and community structure are quite different. It is usually divided into several vegetation subtypes based on the dominant species or leaf characteristics of the community.

Since the desert is in an extremely unbalanced state of water and heat, the ecosystem consists of few species of plants, animals, and microorganisms, and the food chain is relatively simple. Among the animals, there are mainly reptiles like lizards, as well as gerbils, wild asses, argali, ibexes, wild camels, and oryxes. As a very fragile ecosystem, once it disappears, it is difficult to recover.

1.4 Grasslands

Grassland is one of the most important types of terrestrial ecosystems in the world. It is mainly composed of perennial xerophytic grasses and a few shrubs. The world's grasslands are distributed from temperate to tropical. They all occupy a special ecological position, both between the moist forest area and the arid desert area. It is an ecosystem formed under specific semiarid hydrothermal conditions. According to the composition and geographical distribution of grassland, it can be divided into two types, namely, temperate grassland and savanna.

The grasslands in China are mainly temperate grasslands, distributed in the Inner Mongolia Plateau, the western part of the Northeast Plain, the northwestern part of the Loess Plateau, the central part of the Qinghai-Tibet Plateau, and the arid regions in the northwest China. These grasslands are moderate-temperate or cold-temperate grasslands. There are many types of communities, and the species composition and community structure also have certain variations. Grasslands in China are usually divided into several vegetation types. The clumped grassland is the most widely distributed zonal vegetation type, and dominated by moderate-temperate xerophytic cespitose hemicryptophytes, mainly *Stipa* spp., ecologically characterized by adaptations in drought and cold winter climate. According to the difference of ecological and geographical environment, the clumped grassland can also be divided into several vegetation subtypes.

China's savanna area is small, mainly developed in the dry-hot valleys of Yunnan and Sichuan. It is a grassland type with trees and shrubs scattered on the background of mesophytic drought-tolerant grasses.

1.5 Swamp and Aquatic Vegetation

Swamp and aquatic vegetation types are composed mainly of wet plants and aquatic plants that live in conditions with excess moisture in the soil. Three vegetation types are included: woody marsh, herbaceous marsh, and aquatic vegetation (Fig. 3). China's swamps are widely distributed, with the swamp area second only to Canada and the former Soviet Union, with a total area of 14 million hectares, accounting for 1% of China's total land area. They are mainly distributed in the mountainous area of the Northeast, Three River Plain, and Zoige Plateau (including four counties: Zoige, Hongyuan, Aba in Sichuan, and Maqu in Gansu). These three areas are where China's swamps are most concentrated. The source of the Yangtze River in the hinterland of the Qinghai-Tibet Plateau also has a certain area of swamp distribution. Marshes are only sporadically distributed in the subtropical areas. Apart from the middle and lower reaches of the Yangtze River and the major lakes, a small number of swamps are developed in the intermountain basins or in the glacial valleys. In the warm temperate zones, swamps are only found on the edge of rivers and lakes. Chen (2014) believed that mangroves distributed in tropical coastal areas are also woody marshes.

Aquatic vegetation is a type of vegetation that grows in water and consists of aquatic plants. The environment in the water has its own particularity. The variation of water body is on the aspects of depth, fluidity, transparency, light intensity, pH, mineral nutrient content, and oxygen and carbon dioxide comprised. These differences make aquatic plants of different ecological types and have their own adaptation characteristics. For example, submerged plants are born at the bottom of the

Fig. 3 Some swamps and aquatic vegetations in China

water, with the bodies below the water surface, and some of them only expose the flowers out of the surface during the flowering period. Floating plants are those that float on the surface of the water and are further divided into two types: leaf-floating type plants with roots fixed in mud at the bottom of water and full-floating types that have their whole body floating in water. Emergent plants grow upright in water, with the leaves above the water surface, while only the roots, parts of the stems, and petioles inside the water. Usually, aquatic plants are regularly distributed in the water body. Where the water varies from shallow to deep, there are sequenced emergent plant vegetation belts, floating plant vegetation belts, and submerged plant vegetation belts.

2 An Overview of China's Plant Diversity

The native plants of China represent an important part of our global biological heritage. The rich plant species diversity, the high rate of endemism, the presence of many relictual species from the Mesozoic era and Tertiary period, and the representation in the flora of numerous floristic regions, together with a long history of human culture on the land, have resulted China in a botanical richness paralleled in few other parts of the world (Huang et al. 2013).

2.1 Terrestrial Plant Diversity

China is one of the world's mega-diverse countries (Mittermeier et al. 1997), with 36,152 species (18,213 endemic) of terrestrial plant, or higher plant, including 3021 species of mosses, 2147 ferns and fern allies, 262 gymnosperms, and 30,722 angiosperms (Qin et al. 2017), which are, respectively, 23%, 16%, 24%, and 9% of the world's total (Table 1).

China is the home to the world's second richest flora, next to Brazil (Li and Miao 2016). The diversity of extant vascular plants in China is extraordinary, particularly in the northern hemisphere. China, the United States, and Europe have similar areas, but China has more biodiversity than the other two regions: there are more about 33,000 vascular plants (including ferns and seed plants) in China, in contrast to 19,000 in the United States and 11,500 in Europe (Raven 2011), making the ratio of numbers of species in the three areas approximately 3 to 2 to 1. There are a number of factors to account for these differences in diversity, but the most important must be that China includes a significant area of moist tropical and subtropical regions, almost absent in the United States and totally absent in Europe (Zhang and Gilbert 2015).

The abundance of plant species can also be expressed in terms of the size of the family and genus, that is, the number of species they contain. The world's five largest families, which contain more than 10,000 species, are Asteraceae, Fabaceae, Orchidaceae, Poaceae, and Rubiaceae. The first four families are also the top four families of Chinese seed plants, each containing 2577, 2166, 1514, and 2148 species, respectively. The fifth family in China's flora is Rosaceae, with 1461 species, and the other families having more than 1000 species are Ranunculaceae (1229 species), Lamiaceae (1189 species), and Ericaceae (1021 species) (Qin et al. 2017).

The top ten large genera in the angiosperm are *Rhododendron* (650 species), *Carex* (620 species), *Astragalus* (450 species), *Pedicularis* (372 species), *Corydalis* (360 species), *Saussurea* (340 species), *Primula* (300 species), *Elatostema* (300 species), *Salix* (275 species), and *Gentiana* (265).

About one-third of more than 33,000 vascular plants in China, or 11,000 plant species, are woody plants, including more than 3200 species of trees (Fang et al. 2009). For example, China has almost all genera of temperate woody plants, including *Acer* (99 species), *Betula* (40 species), *Carpinus* (36 species), *Fraxinus* (27 species), *Salix* (275 species), *Sorbus* (101 species), etc. Especially central China, it is the area with most abundant deciduous woody plants in the world.

2.2 Endemism

Endemism is a phenomenon in which a biota (such as a species, genus, or family) is confined within a geographic region without exceeding that region. It is one of the

Table 1 A comparison of global and Chinese plant diversity[a, 1]

	Family number			Genus number			Species number		
Group	China	World	China%	China	World	China%	China	World	World%
Mosses	150	215	70	591	1254	47	3021	12,800[2]	23
Ferns and fern allies	40	71	56	178	381	46	2147	13,300[3]	16
Gymnosperms	10	15	67	45	79	57	262	1090[4]	24
Angiosperms	264	~400	66	3187	~10,000	32	30,722	369,000[5]~	8
Total	464			4001			36,152	396,190	9

[a]Data sources: (1) Qin et al. 2017 and Catalogue of Life China (CoL-China, http://sp2000.org.cn/); (2) Crosby et al. 2000; (3) Hassler 2016; (4) WCSP 2016; (5) Kew 2016

Table 2 Number of endemic species and genera of China's higher plants[a]

Group	Endemic/total spp.	Endemic spp. (%)	Endemic/total genera	Endemic gen. (%)
Mosses	553/3021	18	7/591	1.18
Ferns and fern allies	839/2147	39	3/178	1.69
Gymnosperms	148/262	56	5/45	11.1
Angiosperms	16,673/30,722	54	189/3187	5.93
Total	18,213/36,152	50	204/4001	5.10

[a]Data sources: Qin et al. 2017

indicators for measuring the quality of a region's diversity. Areas with high endemism are often areas with abundant biodiversity. Therefore, the study of endemic phenomena is of great significance for the identification of biodiversity hot spots and priority protected taxa and areas (Kier et al. 2009; Huang et al. 2014).

Due to the vast territory, the diversity of landscapes, the diversity of flora, the habitat, and the richness of its constituent elements, the endemism of plants in China is very obvious. Table 2 shows the uniqueness of different groups of higher plants in China. It can be seen that gymnosperms have the highest endemism in genus and species, while mosses are the lowest (Yang 2015).

There are 18,213 endemic species in China's flora, accounting for 50% of the total number of 36,152 (Table 2). If only vascular plants (excluding mosses) are counted, the Chinese plant species endemic rate is about 53% (17,660/33,131). This endemic ratio is much higher than any reported northern hemisphere temperate countries, only lower than Australia and Papua New Guinea (both 91%), Madagascar (74%), South Africa (65%), and Brazil (56%) (Giam et al. 2010). Most of these countries and regions are island-type countries or territories composed of archipelagic islands, mostly located in the southern hemisphere tropical, with a high degree of specificity.

There are two endemic families in China: the Ginkgoaceae of gymnosperms and the Eucommiaceae of angiosperms. In addition, the following six monotypic families are mainly distributed in China, i.e., Bretschneideraceae, Rhoipteleaceae, Tetracentraceae, Cercidiphyllaceae, Trochodendraceae, and Circaeasteraceae, with only a small portion of their distribution range across neighboring countries (Wu and Wang 1983).

There are 204 endemic genera of Chinese higher plants, including seven genera of mosses, three genera of ferns and fern allies, five genera of gymnosperms, and 189 genera of angiosperms. The endemic genera of the first three groups are all monotypic genera, i.e., only one species in one genus. The endemic genus of angiosperms is distributed at 64 families, with 31 families having only one genus. The top five large families with the highest number of endemic genera are Asteraceae (18 genera), Gesneriaceae (15 genera), Orchidaceae (14 genera), Apiaceae (11 genera), and Brassicaceae (9 genera). All are in herbaceous habit (Table 3).

Table 3 Name list of China's endemic higher plant genera (except angiosperms)

Mosses
Brachymeniopsis Broth. (Funariaceae)
Juratzkaeella W. R. Buck (Brachytheciaceae)
Leptocladium Broth. (Leskeaceae)
Pseudopterobryum Broth. (Pterobryaceae)
Scabridens E. B. Bartram (Leucodontaceae)
Sciaromiopsis Broth. (Amblystegiaceae)
Yunnanobryon Shevock, Ochyra, He & Long (Regmatodontaceae)
Ferns and fern allies
Craspedosorus Ching & W. M. Chu (Thelypteridaceae)
Cystoathyrium Ching (Cystopteridaceae)
Neocheiropteris Christ (Polypodiaceae)
Gymnosperms
Cathaya Chun et Kuang (Pinaceae)
Ginkgo L. (Ginkgoaceae)
Metasequoia Miki ex Hu et Cheng (Taxodiaceae)
Pseudolarix Gord. (Pinaceae)
Pseudotaxus Cheng (Taxaceae)

Data sources: Qin et al. (2017)

In addition, there are many genera that are mainly distributed in China, or China is the distribution center of these genera. The most famous are *Rhododendron*, 650 species in China (650/1000, accounting for 65%), *Saussurea* (3401/440, 77%), *Pedicularis* (372/600, 62%), *Camellia* (102/130, 78.5%), *Primula* (300/500, 60%), *Gentiana* (265/380, 70%), *Aconitum* (220/400, 55%), *Corydalis* (360/465, 77%), and *Acer* (99/129, 77%).

2.3 Rich Primitive Taxa

Because of its unique geological history, China has long maintained favorable conditions for plant survival and reproduction. For this reason, a large number of ancient and primitive plant families and genera are found, and many of these relict species have been preserved in China.

The gymnosperm families of the Cycadaceae, Ephedraceae, Ginkgoaceae, and Gnetaceae are rather isolated phylogenetically. Of the eight families of the extant pines and cypresses in the world, all but the Araucariaceae and Sciadopityaceae can be found natively in China. Some world-known conifer species are only native, endemic, as well as relictual to China, such as *Metasequoia glyptostroboides, Glyptostrobus pensilis*, *Cathaya argyrophylla*, *Amentotaxus argotaenia,* and *Pseudotaxus chienii*.

Angiosperms in China include primitive, intermediate, and advanced groups, in terms of evolution. The Magnoliaceae, for example, is commonly recognized as the most primitive angiosperm family. Of the existing 300 species in 17 genera, China has 112 species in 13 genera, including *Liriodendron,* the well-known relict of the Tertiary period. The Hamamelidaceae is another ancient and complex family, with more than 140 species and 30 genera worldwide. The center of its distribution is in China, where there are more than 74 species in 18 genera. Among them, the most primitive genera, namely, *Disanthus*, *Exbucklandia*, and *Rhodoleia*, can all be found in south China (Zhang and Lu 1999).

Thus, the Chinese flora is widely considered to be of primitive origin and rich in major components viewed as descendants or relicts of the paleotropics of the Tertiary period (Axelrod et al. 1998; Wu and Wang 1983).

2.4 Flora of China and the Catalogue of Life China

Since the 1990s, China's plant taxonomists and biodiversity researchers have made major breakthroughs in understanding and mastering the national plant diversity background and its dynamics. This is mainly due to the completion of the two largest floristic projects in the world, the Chinese- language *Reipublicae Popularis Sinicae* (FRPS) (Editorial Committee of FRPS 1959–2004) and the English-language *Flora of China* (FOC) (Wu et al. 1994–2013), as well as *the Catalogue of Life China* (Col-China, plant part) (Qin et al. 2017).

The FRPS includes 80 volumes in 126 books, totaling more than 50 million words. It documented 31,142 species of vascular plants in 3402 genera and 301 families (Yang et al. 2005). It was completed and edited by 312 plant taxonomists and 164 scientific artists from more than 80 institutes and universities across the country (Table 4). It took 45 years to complete. FRPS is currently the largest and most abundant flora in the world and the first encyclopedia of plant resources in China. The publication of the FRPS is of great significance to the research of Chinese botany and related disciplines, the protection of biodiversity, and the sustainable use of resources in China.

The FOC is a revised edition of the FRPS. It is the first English version of a Chinese flora that was completed through the cooperation between Chinese scholars and foreign scholars. The FOC reflects the latest developments and achievements of plant taxonomy in China and also makes up for the inadequacy of reference to

Table 4 Chinese's flora and name list

Project	Volume	Time period covered	Family/genus/species (vascular plant)	Remark
FRPS	80 vols, 126	1959–2004	301:3408:31,142	
FOC	25 vols	1994–2013	312:3328:31,362	
CoL-China	12 vols	2013–2018	314:3410:33,131	Bryophyte 150:591:3021

literatures and type specimens in compilation of FRPS. The FOC project published 49 volumes, including 24 volumes of text, 24 volumes of illustration, and 1 volume of general introduction, documenting 31,362 species of vascular plants of 3328 genera and 312 families, of which 2129 species are ferns and fern allies, 237 species are gymnosperms, and 28,996 species are angiosperms (Wu et al. 2013) (Table 4). The FOC is also very important to introduce Chinese biodiversity to the world. It took 25 years to make the publication, with contributions from 478 botanical experts from 30 countries and 324 artists (scientific painters).

For a mega-diverse country like China to complete two sets of national flora at almost the same time is unique in the world. This is the result of long-term research and accumulation by the Chinese botanical community and the support of the Chinese government to scientific research on biodiversity and profit from the current rapid access to information resources, as well as major breakthroughs brought about by research methods such as cladistics and molecular systematics.

The Catalogue of Life China (CoL-China) was established to compensate the shortcomings of flora, such as long publishing cycle and the lack of newest taxonomical updates. CoL-China may meet timely the needs of biological cataloging for biodiversity monitoring and conservation. Since 2008, the Chinese Academy of Sciences (CAS) has organized a number of taxonomic experts, with reference to the latest taxonomical literature, according to the Species 2000 dataset, to catalogue described species in China and to publish the catalogue/checklist electronically every year, including the Internet version (http://www.sp2000.org.cn/) and CD-ROM versions. The annual checklist covers species names, literature, and distribution of animals, plants, and fungi in China. Therefore, China is the first country in the world to publish the national annual checklist continuously.

The 2017 annual checklist of CoL-China of plant recorded 36,152 species, 4001 genera, and 464 families of terrestrial plant (higher plant), including 1276 important cultivated plants and naturalized plants (Qin et al. 2017).

2.5 *Number of New Taxa Published Annually*

In the past 20 years, China's botanists have strengthened the investigation and collection of gap areas of survey and groups. More and more groups were revised. Therefore, many new taxa and new records have been published annually.

A total of 4213 new species of Chinese plants (including new records and new names, the same below) were published between 2000 and 2019, belonging to 919 genera and 218 families. There are 135 species of mosses, 267 species of ferns and fern allies, 25 species of gymnosperms, and 3786 species of angiosperms (Table 5). According to the statistics, 278 new species were published in 2007, which is highest, whereas the lowest 150 species in 2004 (Fig. 4), an average of 212 new species per year. Globally, 2000 new species are described for average each year (Prance et al. 2000; Kew 2016). Therefore, the number of new species published in China accounts for about one-tenth of the world, and China becomes

Table 5 Newly published species, 2000–2019[a]

	Angiosperms	Gymnosperms	Fern and fern allies	Mosses	Total
Family	140	6	20	52	218
Genus	766	18	57	78	919
Species	3786	25	267	135	4213

[a]Data source: Wu et al. 2013 and Catalogue of Life China (CoL-China), (http://sp2000.org.cn/)

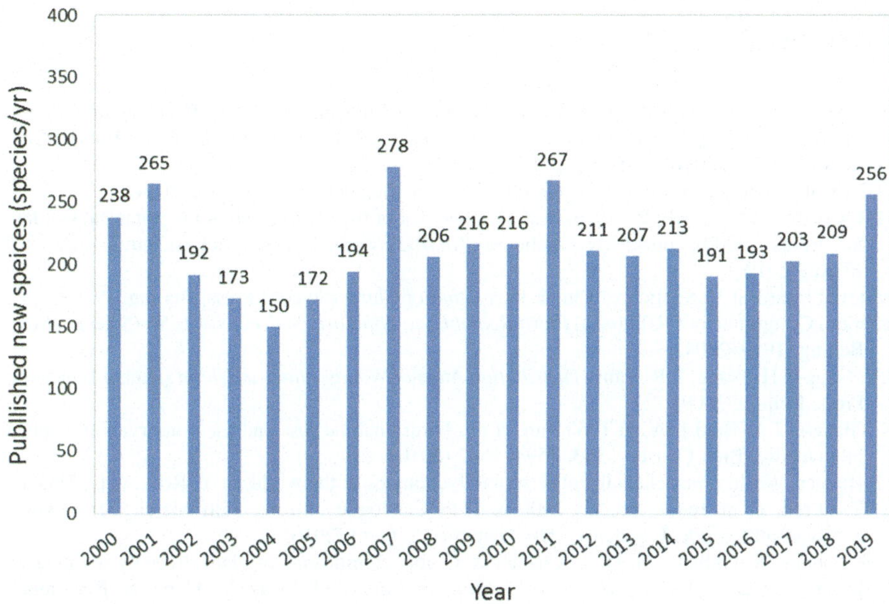

Fig. 4 Published new species, 2000–2019[a]. [a]Data source: Wu et al. 2013 and Catalogue of Life China (CoL-China, (http://sp2000.org.cn/))

one of the top three source countries, besides Australia and Brazil, for the new species of vascular plants entered into IPNI (the International Plant Names Index (http://www.ipni.org) over the past decade (Kew 2016).

There are six families that hold 200 or more new species. They are Orchidaceae (382 species), Asteraceae (251 species), Urticaceae (244 species), Gesneriaceae (241 species), Rosaceae (203 species), and Fabaceae (202 species).

In the past 20 years, at least three new families and 62 new genera have been published from China flora. The new families are Borthwickiaceae (Su et al. 2012), Wightiaceae (Liu et al. 2019), and Pteridryaceae (Zhou et al. 2018). These new genera are distributed in 27 families, including Orchidaceae (9 genera), Asteraceae and Brassicaceae (8 genera for each), Gesneriaceae and Fabaceae (5 genera for each), and Campanulaceae and Magnoliaceae (3 genera for each). Another 21 families have 1–2 genera.

The discovering of new species to China flora is going on every year. The species number of "waiting list" is different from author to author(s), from 2000 (Raven

2011) to 5334 (Lu and He 2017). The discovery journey maybe last for the next 50 years should the current discovery rate per year persist (Lu and He 2017). It's reasonable to estimate that almost all of the newly described species are with narrow distribution area, a sparse number of individuals, and of conservation concern when they are found.

References

D.I. Axelrod, I. Al-Shehbaz, P.H. Raven, History of the modern flora of China, in *Floristic Characteristics and Diversity of East Asian Plants*, ed. by A. L. Zhang, S. G. Wu, (Higher Education Press, Beijing, 1998), p. 12

L.Z. Chen, *China's Flora and Vegetation Geography* (Science Press, Beijing, 2014)

M.R. Crosby, R.E. Magill, B. Allen, et al., A Checklist of the Mosses (Missouri Botanical Garden, St. Louis, 2000), http://www.mobot.org/MOBOT/tropicos/most/checklist.shtml. Accessed 17 Aug 2019

Editorial Board of Vegetation of China, *Vegetation of China* (Science Press, Beijing, 1980)

Editorial Committee of FRPS (ed.), *Flora Reipublicae Popularis Sinicae*, vol 1–80 (Science Press, Beijing, 1959–2004)

J.Y. Fang, Z.H. Wang, Z.R. Tang, *Distribution Atlas of Woody Plants in China* (Higher Education Press, Beijing, 2009)

X. Giam, C.J.A. Bradshaw, H.T.W. Tan, et al., Future habitat loss and the conservation of plant biodiversity. Biol. Conserv. **143**, 1594–1602 (2010)

M. Hassler, World ferns: checklist of ferns and lycophytes of the world, in Y. Roskov, L. Abucay, T. Orrell, et al. (eds). Species 2000 & ITIS Catalogue of Life, 25th March 2016 [www.catalogueoflife.org/col. Species 2000: Naturalis, Leiden] (2016)

H.W. Huang, S. Oldfield, H. Qian, Chapter 2. Global significance of plant diversity in china, in *Plants of China, A Companion to the Flora of China*, ed. by D. Y. Hong, S. Blackmore, (Sciences Press, Beijing, 2013), pp. 7–34

J.H. Huang, K.P. Ma, B. Chen, *Diversity and Geographical Distribution of Endemic Seed Plants in China* (Higher Education Press, Beijing, 2014)

R.B.G. Kew, *The State of the World's Plants Report* (Royal Botanic Gardens, Kew, 2016)

G.H. Kier, T.M. Kreft, W. Lee, et al., A global assessment of endemism and species richness across island and mainland regions. PNAS **106**(23), 9322–9327 (2009)

C.X. Li, X.Y. Miao, Notes on the rank of China in the world in terms of higher plant diversity. Biodivers. Sci. **24**(6), 725–727 (2016)

B. Liu, Y.H. Tan, S. Liu, R.G. Olmstead, D.Z. Min, Z.D. Chen, N. Joshee, B.N. Vaidya, R.C.K. Chung, B. Li, Phylogenetic relationships of Cyrtandromoea and Wightia revisited: A new tribe in Phrymaceae and a new family in Lamiales. J. Syst. Evol. **58**(1), 1–17 (2019)

M.Y. Lu, F.L. He, Estimating regional species richness: The case of China's vascular plant species. Glob. Ecol. Biogeogr. **26**, 835–845 (2017)

R.A. Mittermeier, P.R. Gill, C.G. Mittermeier, *Megadiversity: Earth's Biologically Wealthiest Nations* (CEMEX, Mexico, Agrupación Sierra Madre, 1997)

G.T. Prance, H. Beentje, J. Dransfield, et al., The tropical Flora remains under collected. Ann. Mo. Bot. Gard. **87**(1), 67–71 (2000)

H.N. Qin, et al. (eds.). China checklist of higher plants, in the Biodiversity Committee of Chinese Academy of Sciences ed. *Catalogue of Life China* 2017 Annual Checklist. CD-ROM (Science Press, Beijing, 2017)

P.H. Raven, Plant conservation in the future: New challenges, new opportunities. Plant Diver. Resour. **33**(1), 1–9 (2011)

J.X. Su, W. Wang, L.B. Zhang, et al., Phylogenetic placement of two enigmatic genera, *Borthwickia* and *Stixis*, based on molecular and pollen data, and the description of a new family of Brassicales, Borthwickiaceae. Taxon **61**(3), 601–611 (2012)

WCSP, World Checklist of Selected Plant Families, 2016. http://apps.kew.org/wcsp

Z.Y. Wu, H.S. Wang, *Chinese Physical Geography—Phytogeography* (Science Press, Beijing, 1983)

Z. Y. Wu, P. H. Raven, D. Y. Hong (eds.), *Flora of China*, vol 2–25 (Science Press, Beijing, 1994–2013)

Z. Y. Wu, P. H. Raven, D. Y. Hong (eds.), *Flora of China*, Introduction, vol 1 (Sciences Press, Beijing, 2013)

Y. Yang, Diversity and geographical distribution of gymnosperms in China. Biodivers. Sci. **23**(2), 243–246 (2015)

Q.E. Yang, G.H. Zhu, D.Y. Hong, Z.Y. Wu, P.H. Raven, World's largest flora completed. Science **309**, 2163 (2005)

L.B. Zhang, M.G. Gilbert, Comparison of classification of vascular plants of China. Taxon **64**(1), 17–26 (2015)

Z.Y. Zhang, A.M. Lu, Hamamelidaceae: Distribution, fossil history and origination, in *Phytogeography of Seed Plant Families and Genera*, ed. by A. M. Lu, (Science Press, Beijing, 1999), pp. 196–217

X.M. Zhou, L. Zhang, N.T. Lu, X.F. Gao, L.B. Zhang, Pteridryaceae: A new fern family of Polypodiineae (Polypodiales) including taxonomic treatments. J. Syst. Evol. **56**(2), 148–173 (2018)

Rare and Endangered Plants in China

Haining Qin, Xiaohua Jin, and Lina Zhao

Abstract During 2008 to 2013, China botanists completed a comprehensive assessment of the threatened status of China flora, using the International Union for Conservation of Nature *(IUCN) Red List Categories and Criteria*, version 3.1, and released the *Red List of China Higher Plants* (RLCHP). Therefore, China becomes the second mega-diversity country, after South Africa, which fully assessed the status of its whole flora, to achieve Target 2 of the Global Strategy for Plant Conservation (GSPC), i.e., "an assessment of the conservation status of all known plant species, as far as possible, to guide conservation action." Herein, we introduced the content of RLCHP, focusing on the threatened species, their components, distributions, threatened factors, etc., and briefly demonstrated how the Red List be used as a conservation tool in China biodiversity research and monitoring. We indicated finally that the Red List is not once for all, which needs to be updated regularly, and it is essential to establish a national infrastructure of the Red List and biodiversity conservation, allowing the Red List knowledge transfer and policy support in the country.

Keywords Conservation assessments · The IUCN Red List criteria · China flora · Threatened species

H. Qin (✉) · L. Zhao
State Key Laboratory of Systematic and Evolutionary Botany, Institute of Botany, Chinese Academy of Sciences, Beijing, China

College of Life Sciences, University of Chinese Academy of Sciences, Beijing, China
e-mail: hainingqin@ibcas.ac.cn; zhaoln@ibcas.ac.cn

X. Jin
State Key Laboratory of Systematic and Evolutionary Botany, Institute of Botany, Chinese Academy of Sciences, Beijing, China
e-mail: xiaohuajin@ibcas.ac.cn

H. Ren (ed.), *Conservation and Reintroduction of Rare and Endangered Plants in China*, https://doi.org/10.1007/978-981-15-5301-1_2

1 Threatened Species Overview

Many countries in the world, even developed countries in Europe and America, are faced with a shortage of conservation funds, and the number of species often far outnumbers the limited conservation resources obtained. Therefore, determining biodiversity conservation strategies and implementing protection measures require both consideration of the overall situation and identifying priorities. In other words, protecting as many species as possible with the limited resources, and protecting the most deserving and most urgently needed species with optimal resources, is the primary responsibility of current biodiversity conservation efforts. In determining the target species that are most worthy of protection, researchers often use certain assessment methods to identify whether the species is in an endangered status, whether it is a rare and endemic group, and whether it is of high genetic value, economic value, and scientific value.

The IUCN Red List of Threatened Species (hereafter, the IUCN Red List), established by IUCN, is currently the most widely used system for assessing the probabilities of extinction for taxa and is also the method of evaluation recommended by the Secretariat of the Convention on Biological Diversity (CBD) (Vié et al. 2008). The aim of the IUCN system is to identify taxa that are facing a high probability of extinction, using objective methods and quantitative criteria, such as population size, distribution range, and decline. Any taxa can be classified into one of the nine categories depending on their probabilities of extinction, i.e., Extinction (EX), Extinct in the Wild (EW), Critically Endangered (CR), Endangered (EN), Vulnerable (VU), Near Threatened (NT), Least Concern (LC), Data Deficient (DD), and Not Evaluated (NA). In a non-global regional assessment, such as the national level plus the regional extinction (RE) level, CR, EN, and VU levels are considered collectively as threatened, and the species are called threatened species. Threatened species are recognized as priority targets for biodiversity conservation.

List of Rare and Endangered Protected Plants in China (SEPA & IB-CAS, 1987) is China's first list of protected plants and the first to adopt the concept of the IUCN Red List. The list includes 388 vascular plants, using three categories, i.e., endangered, rare, and vulnerable, and correspondingly sets them to three protection grades: grade one, two, and three. The subsequent publication of the *China Plant Red Book: Rare and Endangered Plants* (Volume 1) (Fu and Jin 1991) provided information on the morphological description, distribution, habitats, protective measures, etc. Color photos and distribution maps were also provided for each plant. The publication of the list and the book is the pioneering work of endangered and protected plant list in China and lays significant foundations for countrywide plant conservation.

The *China Species Red List*: Vol. I *Red List* (Wang and Xie 2004) is the first Red List in China using the *IUCN Red List Categories and Criteria*, version 3.1. The list/book consists of two parts: the animal list and the (seed) plant list. The plant part is evaluated for 4408 target species, of which 3782 are listed as threatened species, accounting for 86% of the assessed species.

The *Red List of China Higher Plants* (RLCHP) was initiated by the Ministry of Environmental Protection (MEP), in 2008, with the cooperation of Chinese Academy of Sciences (CAS). The global IUCN Red List, version 3.1 (IUCN 2012a, b), and Guidelines for Application at Regional and National Levels (IUCN 2012a, b) were utilized to evaluate all native higher plants on the basis of comprehensive data collections. To guarantee the assessment quality, 300 experts are invited to involve the project which lasted for several years. The RLCHP was officially released on the MEP official website in 2013 (http://www.mee.gov.cn/gkml/hbb/bgg/201309/t20130912_260061.htm). Later, in 2017, the RLCHP was minor revised and published in a special issue by a Chinese journal, *Biodiversity Science* (Qin et al. 2017a, b). The following is an introduction to the RLCHP compiling process, analysis to the results, and some application and recommendation on the plant conservation in China.

2 Introduction to *the Red List of China Higher Plants* (RLCHP)

2.1 Overview of the Red List

The evaluation results of the Red List of 35,784 wild higher plants in China indicated that 21 species are Extinct (EX), nine are Extinct in the Wild (EW), ten species are categorized as Regionally Extinct (RE), 614 species are Critically Endangered (CR), 1313 species are Endangered (EN), 1952 species are Vulnerable (VU), and 2818 species are described as Near Threatened (NT), while 24,243 species are considered Lease Concern (LC) and 4804 species have Data Deficient (DD) (Table 1, Fig. 1). A total of 3879 species may be categorized as threatened species, accounting for

Table 1 Total number of China higher plants' taxa listed in the various Red List categories[a]

Categories	Mosses	Fern and fern allies	Gymnosperms	Angiosperms	Total	Percentage (%)
EX	0	0	0	21	21	0.06
EW	0	0	0	9	9	0.03
RE	0	0	0	10	10	0.03
CR	16	43	37	518	614	1.72
EN	58	68	35	1152	1313	3.67
VU	112	71	76	1693	1952	5.45
NT	214	66	0	2538	2818	7.88
LC	1900	1124	87	21,132	24,243	67.75
DD	921	872	16	2995	4804	13.42
Total	3221	2244	251	30,068	35,784	100

[a]Data sources: http://www.mee.gov.cn/gkml/hbb/bgg/201309/t20130912_260061.htm and Qin et al. (2017a, b)

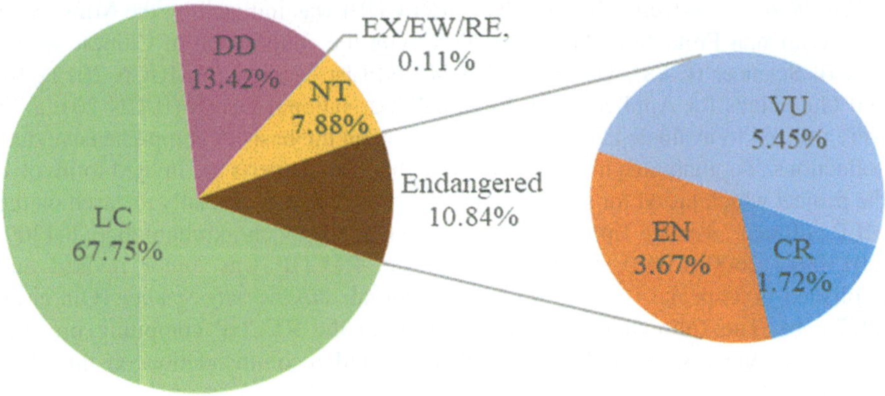

Fig. 1 Taxa classified into *the Red List of China Higher Plants* categories as a proportion of the total flora. **Note:** Endangered species (Threatened species) includes the IUCN categories CR, EN, and VU

10.84% of the total number of species assessed, which are the priority targets for plant protection in China. In addition, the 2818 NT species and 4804 DD species are of conservation significance and should be paid attention to as well.

There are 17,700 endemic species in China, accounting for 49.7% of the total (Fig. 2). These endemic species are great germplasm resources and have genetic diversity values. Among the 3879 threatened species, 2462 are endemic, accounting for 13.9% of the total number of endemic species and 63.5% of the threatened species. It is obvious that endemic species are highly more threatened than non-endemic species. Regarding the unique resource to the country, threatened endemic species should be paid more attention for conservation.

2.2 *The Overview of Extinct Species*

The assessment indicated that 40 species are extinct in total, and all are angiosperms (Table 2). The main threats to the extinct species are the degradation and loss of habitats. The long-term economic activities of human beings have changed the nature of land use, turning wild plant habitats into agricultural land, animal husbandry, and fishery products or urban construction land and road network pipeline sites, etc. Over and over, the number of plant populations has decreased until the species has disappeared. Overuse of economic plants is another main threat to the species extinction. For example, the extinction of the precious medicinal plant *Panax notoginseng* (Araliaceae) and the famous ornamental tree species *Michelia velutina* (Magnoliaceae) is caused by long-term overexploitation.

Fig. 2 Some endemic and rare species in China

2.3 *The Overview of Threatened Species*

2.3.1 General Analysis by Groups

The *Red List of China Higher Plants* (RLCHP) includes a total of 3879 threatened (CR, EN, VU) species, accounting for 10.84% of the total species evaluated, including 186 species of mosses, 182 species of ferns and fern allies, 148 species of gymnosperms, and 3363 species of angiosperms. The group with the highest threat ratio was gymnosperms (51%), followed by angiosperms (11%), ferns and fern allies (6.28%), and mosses (4.57%) (Qin et al. 2017a, b).

Among the gymnosperms, the Cycadaceae has the highest ratio of threatened species; all 22 are threatened species (100%). The cycad plant has been overexploited for a long time due to its high ornamental value. In addition, the plant grows slowly, and its habitat is specialized and easily degenerated. The Taxaceae (yew) is another group with higher threat ratio; 11 species are all threatened (100%). As the source of the cancer-fighting taxanes, the yew is heavily exploited for its medicinal value and material needs.

The angiosperms are the principal category of green plant community on the planet, and the number of endangered angiosperm species is also the largest. Among the angiosperms in China, the family that has the highest ratio of threatened species

Table 2 Extinct species of RLCHP[a]

Categories	Name	Family
EX	*Friesodielsia hainanensis* Tsiang and Li	Annonaceae
EX	*Ligularia parvifolia* Chang	Asteraceae
EX	*Begonia* sublongipes Y. M. Shui	Begoniaceae
EX	*Betula halophila* Ching ex P. C. Li	Betulaceae
EX	*Corsiopsis chinensis* Zhang, Saunders, and Hu	Corsiaceae
EX	*Elaeagnus liuzhouensis* C. Y. Chang	Elaeagnaceae
EX	*Rhododendron xiaoxidongense* W. K. Hu	Ericaceae
EX	*Gyrogyne subaequifolia* W. T. Wang	Gesneriaceae
EX	*Chelonopsis siccanea* W. W. Smith	Lamiaceae
EX	*Ombrocharis dulcis* Hand.-Mazz.	Lamiaceae
EX	*Stauntonia obcordatilimba* C. Y. Wu & S. H. Huang	Lardizabalaceae
EX	*Beilschmiedia ningmingensis* S. Lee & Y. T. Wei	Lauraceae
EX	*Machilus salicoides* S. Lee	Lauraceae
EX	*Lilium stewartianum* Balf. f. & W. W. Sm.	Liliaceae
EX	*Eulophia monantha* W. W. Sm.	Orchidaceae
EX	*Gastrochilus nanchuanensis* Z. H. Tsi	Orchidaceae
EX	*Liparis hensoaensis* Kudo	Orchidaceae
EX	*Tainia emeiensis* (K. Y. Lang) Z. H. Tsi	Orchidaceae
EX	*Lepisanthes unilocularis* Leenh.	Sapindaceae
EX	*Pedicularis humilis* Bonati	Scrophulariaceae
EX	*Premna mekongensis* W. W. Sm. var. meiophylla W. W. Sm.	Verbenaceae
EW	*Panax notoginseng* (Burkill) F. H. Chen ex C. H. Chow	Araliaceae
EW	*Rhododendron adenosum* David.	Ericaceae
EW	*Rhododendron kanehirae* Wils.	Ericaceae
EW	*Chirita spadiciformis* W. T. Wang	Gesneriaceae
EW	*Najas pseudogracillima* Triest	Hydrocharitaceae
EW	*Aspidistra austrosinensis* Y. Wan and C. C. Huang	Liliaceae
EW	*Nymphoides lungtanensis* S. P. Li, T. H. Hsieh and C. C. Lin	Menyanthaceae
EW	*Rhamnus tzekweiensis* Y. L. Chen & P. K. Chou	Rhamnaceae
EW	*Curcuma exigua* N. Liu	Zingiberaceae
RE	*Hydrocera triflora* (L.) Wight. & Arn.	Balsaminaceae
RE	*Dioscorea poilanei* Prain & Burkill	Dioscoreaceae
RE	*Lithocarpus cryptocarpus* A. Camus	Fagaceae
RE	*Myriophyllum tetrandrum* Roxb.	Haloragaceae
RE	*Ottelia cordata* (Wall.) Dandy	Hydrocharitaceae
RE	*Michelia velutina* DC.	Magnoliaceae
RE	*Bulbophyllum yunnanense* Rolfe	Orchidaceae
RE	*Potamogeton alpinus* Balbis	Potamogetonaceae
RE	*Diploknema yunnanensis* Tao, Yang, and Zhang	Sapotaceae
RE	*Premna pyramidata* Wall. ex Schauer	Verbenaceae

[a]Data source: http://www.zhb.gov.cn/gkml/hbb/bgg/201309/t20130912_260061.htm

Table 3 Top 15 angiosperm threatened families with more than 20 species

Family	Threatened species	Total species	Threatened ratio (%)
Magnoliaceae	76	106	71.7
Actinidiaceae	48	72	66.7
Dioscoreaceae	31	57	54.4
Hamamelidaceae	38	75	50.7
Aristolochiaceae	40	81	49.4
Orchidaceae	666	1538	43.3
Buxaceae	10	25	40
Sterculiaceae	30	77	39
Aceraceae	47	129	36.4
Styracaceae	23	65	35.4
Icacinaceae	8	23	34.8
Arecaceae	19	56	33.9
Juglandaceae	7	22	31.8
Elaeocarpaceae	17	55	30.9
Menispermaceae	23	79	29.1

is Magnoliaceae, in which 76 species (71.7% of the total) are threatened, followed by Actinidiaceae, with 48 threatened species, a threat ratio of 66.7%. Magnolias are threatened because of their beautiful tree shape and large flowers, which make them important ornamental trees. The Actinidiaceae (kiwifruit plants) are mainly harmed by human economic activities that cause their habitat to be fragmented or even lost, leading to endangerment. Table 3 shows the highest threat levels in more than 20 families.

2.3.2 Analysis by Area

Southwestern China is abundant of threatened species both at the provincial and county levels. Yunnan province ranges the first, with 1674 species, accounting for 44.4% of the nation's threatened species, accounting for 10.8% of the total number of Yunnan plants, followed by Sichuan, Guangxi, Tibet, Hainan, and Guangdong. See Table 4 for provinces with more threatened species.

2.3.3 Analysis by Altitude

The threatened species are distributed at altitudes from sea level to more than 5000 m, but most concentrated at middle and low altitude areas. Approximately 1302 species are distributed in the zone of 500–1000 m in elevation, 1228 species in the zone of 1000–1500 m, and 991 species in the zone of 1500–2000 m. One possible explanation is that the middle and low altitude areas are densely distributed

Table 4 Provinces with most threatened species[a]

Province	Threatened species	Total species	Threatened/province total (%)	Threatened (%)
Yunnan	1674	15,463	10.8	43.1
Guangxi	644	6894	9.3	16.6
Sichuan	637	10,333	6.2	16.4
Tibet	412	7491	5.5	10.6
Hainan	394	3528	11.2	10.2
Guizhou	378	5694	6.6	9.7

[a]The total number of species in each province is from the basic list of this assessment, for reference only

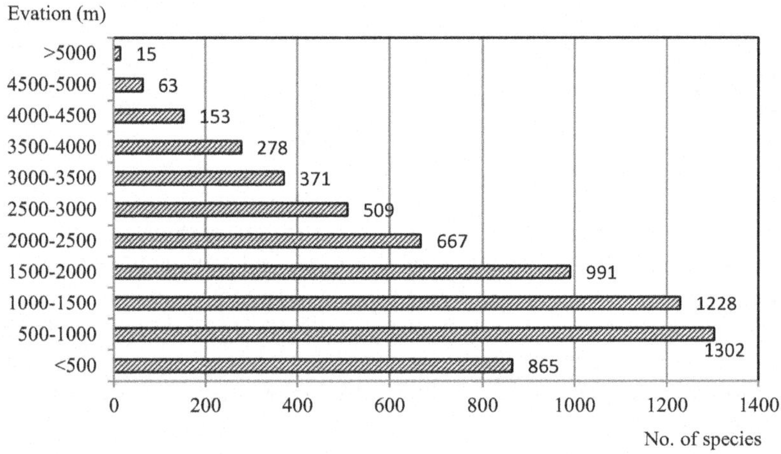

Fig. 3 Threatened species ranged by altitude belts. **Note:** Same species is repeated in different elevation zones

areas of biological species and also the areas with the much more human activities, therefore great disturbance on the plants (Fig. 3).

2.3.4 Analysis by Threat Factors

The analysis of threat factors is the focus and main content of threatened species analysis, which will improve our understanding the causes of species extinction risks and provide accurate methods and countermeasures for the conservation.

RLCHP threat analysis of the threatened species (CR, EN, and VU) is followed the IUCN classification scheme roughly, i.e., (1) habitat degradation and loss, (2) overexploitation for economic use, (3) species intrinsic factors, (4) interspecies effects (including invasive species), (5) environmental pollution, and (6) natural disasters and climate change. However, some specific conditions and characteristics of threats to Chinese plants were also reflected in the analysis. Table 5 shows the

Table 5 Threats to China angiosperms' (only 3525 species which have clear factors)

Threats	Involved species	In total (%)
Habitat degradation and loss	2116	55
Direct mining or felling	963	25
Environmental pollution	17	0.4
Natural disasters and climate change	39	1
Intrinsic factors	357	9
Interspecies effect	33	0.9
Total	3525	

status and proportion of main impact factors in the threatened species of angiosperms.

China's habitat degradation and loss, similar to many countries and regions, are the first factors that cause plants to face the risk of extinction, which is the main threat for 2116 species of angiosperms, accounting for 55% of the total species. Among them, agricultural activities such as land reclamation, afforestation (including economic crops), and overgrazing have the greatest impact on plant habitats, accounting for 18.1%, and infrastructure construction such as road network pipelines, water diversion, damming, and residents' settlements accounted for 7.1%. Resources development activities such as mining quarry accounted for 3%.

Overexploration ranked second in the threats as the main cause of 963 plant endangerment, accounting for 25%, mainly for excessive use of wild resource plants, such as felling of timber species, medicinal plant collection, and transplantation of ornamental plants. This ratio is higher in China than that in other countries. The third risk factor comes from the species' own characteristics, including low seed setting rate and seedling survival rate, slow growth, low population density, and weak diffusion ability, involving 357 species, accounting for 9%. The information on these species is mainly from expert surveys, which also shows the importance of expert extensive participation in the *Red List* assessment.

In contrast, natural disasters and climate change, interspecies effect, and environmental pollution were ranked as fourth, fifth, and sixth, respectively, all at or below 1%. This indicates that the impact of environmental climate and invasive alien species on China's plants is less concerned in the past, and this subject should be one of the new directions and key points of research in China in the future. Only in this way can we fully uncover the health status of the whole Chinese flora and can provide more scientific information for conservation decision makers.

2.4 *Significance of the Assessment*

The completion of the *Red List of China Higher Plants* (RLCHP) provides strong support for the government's implementation of the Convention on Biological Diversity (CBD) (Editorial Committee 2008). More importantly, the list has enabled us to basically understand the endangered status and background of Chinese flora, which will provide an important tool and reference data for the conservation and sustainable use of national and local biodiversity. Since 2013 when the RLCHP was officially released on the MEP website and published in Biodiversity Science in 2017, it has provided decision makers with basic data for biodiversity conservation policies at national and regional level, as well as the important basis for the research teams and individuals to study China's biodiversity conservation area gaps and priority protection (Zhang et al. 2015).

RLCHP is the first rare and endangered plant list that includes quantitative analyses of the threats impacting China plants, and the analyses are done at national, biome, and provincial levels and allow conservation authorities to identify the main causes of loss to plant diversity within their regions. Thus, China becomes the second mega-diversity country, after South Africa, in the world to assess the conservation status of whole flora (Raimondo 2011).

Threatened status assessment of plant species is not once for all. At one hand, following ecological construction of China, increasing protection awareness of the publics and governing bodies, natural habitats will be greatly improved in the foreseeable future. At another hand, utilization intensity of some economically valuable wild species may increase or decrease as the human consumption concept changes or the artificial cultivation method improves. The existence of these subjective and objective factors means that the threatened condition and information acquisition of the species are in the process of dynamic change. Thus, the *Red List of China Higher Plants* (RLCHP) needs to be continuously updated in time. It is necessary for the government to provide continuous special funding for the assessment of plant conservation status; to carry out real-time dynamic monitoring and investigation of wild species with lack of data, habitat changes, and changes in human utilization; and ultimately to achieve an all-information, accurate, and dynamic background for threatened and data-deficient species. Only in this way can we truly know the endangered status and extinction risk of the species and can the assessment really play the role of a conservation tool and provide a scientific and reliable basis for biodiversity monitoring, protection, and implementation of CBD at national and international levels.

Acknowledgments This chapter was modified from the papers published by our group in *Biodiversity Science* (Qin and Zhao 2017; 25:689–695; Qin et al. 2017a; 25:696–744; Qin et al. 2017b; 25: 745–757). The related contents are reused with the permission.

References

Editorial Committee, *China's Strategy for Plant Conservation* (Guangdong Science and Technology Publishing House, Guangzhou, 2008)

L.G. Fu, J.M. Jin, *Chinese Plant Red Book—Rare and Endangered Plants*, vol 1 (Science Press, Beijing, 1991)

IUCN, *IUCN Red List Categories and Criteria, Version 3.1*, 2nd edn. (IUCN, Gland, 2012a)

IUCN, *Guidelines for Application of IUCN Red List Criteria at Regional and National Levels, Version 4.0* (IUCN, Gland, 2012b)

Ministry of Environmental Protection (MEP) and Chinese Academy of Sciences (CAS), Red List of Higher Plants in China, 2013. http://www.zhb.gov.cn/gkml/hbb/bgg/201309/t20130912_260061.htm

H. Qin, L.N. Zhao, Evaluating the threat status of higher plants in China. Biodivers. Sci. **25**(7), 689–695 (2017)

H. Qin, Y. Yang, S. Dong, et al., Threatened species list of China's higher plants. Biodivers. Sci. **25** (7), 696–744 (2017a)

H. Qin, L.N. Zhao, S.X. Yu, et al., Evaluating the endangerment status of China's angiosperms through the red list assessment. Biodivers. Sci. **25**(7), 745–757 (2017b)

D. Raimondo, The Red List of South African plants—A global first. S. Afr. J. Sci **107**(3/4), 2 (2011). https://doi.org/10.4102/sajs.v107i3/4.653

State Environmental Protection Administration (SEPA), Institute of Botany of Chinese Academy of Sciences (IB-CAS), *List of Rare and Endangered Protected Plants in China*, vol 1 (Science Press, Beijing, 1987)

J.C. Vié, C. Hilton-Taylor, C. Pollock, et al., The IUCN red list: A key conservation tool, in *The 2008 Review of the IUCN Red List of Threatened Species*, ed. by J. C. Vié, C. Hilton-Taylor, S. N. Stuart, (IUCN, Gland, 2008)

S. Wang, Y. Xie, *China Species Red List*, Red List, vol I (Higher Education Press, Beijing, 2004)

Z.K. Zhang, J.S. He, J.S. Li, et al., Distribution and conservation of threatened plants in China. Biol. Conserv. **192**, 454–460 (2015)

In Situ and Ex Situ Conservation of Chinese Plants

Xiaohua Jin and Hai Ren

Abstract In total, 28,385 higher plants were distributed in the nature reserves, accounting for 77.0% of the total number of higher plants in China. Among them, there are 3842 species which are effectively protected; 2478 species are better protected. More than 84.4% endangered species grow in national and/or provincial natural reserves. On the national scale, there are 237 state key protected species distributed in national nature reserves (NNRs), approximately 80% of the total key protected plants. More than 30% species with extremely small populations occur in NNRs and provincial nature reserves (PNR), while 35 species (29% of the total) are not covered by any national or provincial nature reserves. There were 162 botanical gardens or arboreta in China. In total, 22,104 native species from 2911 genera and 288 families are cultivated ex situ at botanical gardens or arboreta. These account for 65%, 86%, and 91% of all native flora species, genera, and families, respectively. The number of endangered and threatened plants in ex situ conservation was about 1500 species, which was about 39% of the endangered and threatened species recorded in China. By 2015, Chinese arboreta and botanical gardens had successfully introduced approximately 270 species on the checklist of state key protected wild plant species.

Keywords Natural reserves · In situ conservation · Ex situ conservation · Living collection · Botanical garden

X. Jin (✉)
State Key Laboratory of Systematic and Evolutionary Botany, Institute of Botany, Chinese Academy of Sciences, Beijing, China
e-mail: xiaohuajin@ibcas.ac.cn

H. Ren
South China Botanical Garden, Chinese Academy of Sciences, Guangzhou, China
e-mail: renhai@scbg.ac.cn

H. Ren (ed.), *Conservation and Reintroduction of Rare and Endangered Plants in China*, https://doi.org/10.1007/978-981-15-5301-1_3

1 Introduction

Due to the environmental consequences of population growth and economic development, many plant species are threatened with extinction, resulting in habitat reduction, fragmentation and over-collection, invasion of alien species, climate change, and pollution (Huang 2011; Miller et al. 2012; Ren et al. 2014). Simultaneously, some species are threatened by their own nature, such as reproductive disorders (generally characterized as low fruit set and low germination rate), low genetic diversity, etc. It is estimated that around 94,000–194,000 species are on the verge of extinction in their habitat (Miller et al. 2012), including 3879 plant species in China (Qin and Zhao 2017; Qin et al. 2017; due to insufficient data, many species have not been evaluated). The greatest challenge facing the international community is the conservation, restoration, and sustainable use of plant diversity.

In 2016, there are approximately 2740 national, provincial, and county nature reserves in China with a total area of 1.47 million km^2, about 14.8% of the national land area. Out of which, 446 NNRs account for approximately one million km^2. The network of nature reserves has been well established, including nearly full complete vegetation types of China, a reasonable spatial pattern and diverse ecological functions.

1.1 Distribution of Threatened Species in China

Among the 36,866 higher plants in China, except for the 1929 species that were not evaluated in Hong Kong, Macao, and Taiwan, 28,385 higher plants (77%) were distributed in nature reserves, which received a certain degree of in situ conservation.

Threatened species are widely distributed in China, especially in Yunnan, the Himalayas region, and the Hengduan Mountains, together with Hainan and Taiwan islands. South China (Guangxi, Guangdong, and Fujian) also has a high degree of diversity. Instead, east China (Anhui, Jiangsu, Shandong and Shanghai), northern China (Beijing, Hebei, Inner Mongolia, Shanxi, Tianjin), northeast China (Heilongjiang, Liaoning, Jilin), and northwest China (cold and dry land of Tibet and northwestern Xinjiang) have a relatively low diversity of threatened species. In some parts of Chongqing, Hebei, Jiangsu, and Shandong, there is no record of threatened species. Southern and northern parts of Yunnan province and the southwestern part of Hainan Island have high species richness of China's endemic threatened plants. For different categories, many critically endangered (CR) species are distributed in southwestern China (mainly in Yunnan, Guangxi, Sichuan, and Hunan), but there are less than 20 species; EN species are abundant in southern Yunnan and southwestern Hainan Island. The diversity of VU species is similar to that of the entire threatened species, mainly in southwestern China (mainly in Yunnan, southeastern Tibet, and western Sichuan), Hainan, and Taiwanese mountains (Zhang et al. 2015b).

1.2 The Distribution of Threatened Species in the NNRs and PNRs in China

On average, approximately 18.8% of threatened species occupied areas is in national nature reserves (NNRs) (Fig. 1) (Zhang et al. 2015b). The protective coverage of the narrowest distributed threatened plants (mean = 21.6%) is higher than that of narrow distributed (20.5%), widespread (18.3%), and the most widespread (14.9%) species. However, about 827 threatened species have not been covered in any NNR, of which 19 are the most widely distributed, 125 widely distributed, 260 narrowly distributed, and 423 are the most narrowly distributed (Zhang et al. 2015b). Among the unprotected threatened species, 627 are endemic to China. The NNRs of all CR, EN, and VU species accounted for 18.5%, 20.3%, and 17.9%, respectively. It appears that 183 CR, 263 EN, and 381 VU species have not been covered by any NNR (Zhang et al. 2015b).

In total, NNRs and PNRs together cover 27.5% of all threatened species in China (std = 22%). There are a total of 397 threatened species (including 293 endemic species) covered by neither NNRs nor provincial nature reserves (PNRs), including four most widely distributed, 30 widely distributed, 107 narrowly distributed, and 256 the most narrowly distributed threatened plant (Zhang et al. 2015b). The NNRs and PNRs accounted for 27.0% (std = 25%), 28.7% (std = 22%), and 26.8% (std = 20%) of China's CR, EN, and VU species occupied areas, respectively. Among these species, 93 are CR, 121 are EN, and 183 are VU. Most of the threatened species that are not covered in NNRs are distributed in southwestern China (the borders of Guizhou and Guangxi, southern Tibet, the border regions Yunnan and Sichuan, the southwestern Yunnan), some of which are distributed in the southeast China (mainly the coast regions between Guangdong and Fujian,

Fig. 1 Dinghushan national nature reserve in situ conserves 22 national key protection wild plants

northern Hainan Island, central part of Hubei) and border regions of Xinjiang. When considering both PNRs and NNRs, unprotected threatened plants are mainly distributed in southwestern China (mainly distributed in Yunnan, Guizhou, and Guangxi provinces), central Hainan, and eastern Zhejiang (Zhang et al. 2015b).

2 In Situ Conservation

2.1 Evaluation Overview

According to the number of protected areas with distribution records, the level of in situ conservation of higher plants is divided into seven levels: effective protection, better protection, general protection, less protection, unclear protection status, unprotected, and unevaluated.

2.1.1 Effective Protection

Some wild plants are distributed in more than 30 nature reserves, or the distribution of wild populations is narrow. Although the number of protected areas is less than 30, more than two-thirds of them are distributed in protected areas (only for the very small-population species whose wild population is known that the population number can be quantitatively evaluated, as below).

2.1.2 Good Protection

Some higher plants are distributed in 16–30 nature reserves, or the wild population distribution area of the species is small, but half to two-thirds of populations are distributed in the protected area.

2.1.3 Adequate Protection

Some higher plants are distributed in 6–15 nature reserves.

2.1.4 Less Protection

Some higher plants are only distributed in 1–5 nature reserves.

2.1.5 Uncertain Status

No distribution of the plant in the nature reserve. The division is indeed not protected by in situ or should be distributed within the nature reserve, but there are no records of the existing information.

2.1.6 Unprotected

A certain plant distribution area is relatively narrow and existing data proves that it is not distributed in the protected area.

2.1.7 Not Evaluated

Some higher plants are only distributed in Hong Kong, Macao, and Taiwan regions of China, and there is no distribution record in the mainland; or the species is introduced and cultivated and occasionally recorded in China.

2.2 Protection for Chinese Higher Plants

The evaluation results (Table 1) show that among the 36,866 higher plants in China, except for the 1929 species that were not evaluated in Hong Kong, Macao, and Taiwan, 28,385 higher plants were distributed in nature reserves, and a certain degree of in situ conservation was obtained (Zhang et al. 2015b). Approximately 77.00% of higher plants are in situ conservation. There are 6552 species of higher plants in nature reserves with unclear and unprotected status, accounting for 17.77% of the total number of higher plants in China. Among them, there are 3842 species of higher plants that are effectively protected, or 10.42% of the total number of higher plants in China; 2478 species of higher plants are better protected, or 6.72% in China. There are 16,808 species of higher plants with less protection, 45.59% of the total number of higher plants, and 3115 species of unclear protection status, about 8.45% of higher plants in China. The number of unprotected higher plants was 3437, 9.32% of higher plants in China; and the number of unevaluated higher plants was 1929, 5.23% of higher plants (Zhang et al. 2015b).

2.3 Protection for State Key Protected Plants in China

The checklist of China's key protected plants was announced by the State Forest Department (Yu 1999). On the national scale, there are 237 key protected plants

Table 1 Evaluation of in situ conservation of higher plant species in China

Protection level	Higher plants		Angiosperms		Gymnosperms		Ferns		Bryophytes	
	Number	Ratio[a] %	Number	Ratio %	Number	Ratio %	Number	Ratio %	Number	Ratio %
Effective protection	3842	10.42	3292	10.56	67	18.87	398	14.31	85	3.31
Well protected	2478	6.72	2082	6.68	42	11.83	166	5.97	188	7.32
General protection	5257	14.26	4457	14.30	55	15.49	362	13.01	383	14.91
Less protected	16,808	45.59	14,381	46.15	141	39.72	1309	47.05	977	38.05
Uncertain	3115	8.45	2500	8.02	10	2.82	253	9.09	352	13.71
Unprotected	3437	9.32	2800	8.99	12	3.38	135	4.85	490	19.08
Not evaluated	1929	5.23	1649	5.29	28	7.89	159	5.72	93	3.62
Total	35,866	100	31,161	100	355	100	2782	100	2568	100

[a]Proportion of the higher plants in China

Table 2 The number of state key protected wild plants in the NNRs and the percentage of the total protected wild plants

		Number of families (%)	Number of genera (%)	Number of species (%)	Species under Class I
Ferns		11 (78.57)	14 (82.35)	26 (72.22)	3 (50.00)
Gymnosperms		7 (100.00)	22 (95.65)	56 (84.85)	27 (79.41)
Angiosperms		61 (89.71)	124 (82.12)	155 (79.90)	26 (78.79)
Total		79 (88.76)	160 (83.77)	237 (80.07)	56 (76.71)
Not evaluated	1929 (5.23)	1649 (5.29)	28 (7.29)	159 (5.72)	93 (3.62)
Total	36,866 (100)	31,161 (100)	355 (100)	2782 (100)	2568 (100)

distributed in NNRs, accounting for 80.07% of the total number of key protected plants; 56 of them are Class I protected plants, 76.71% of Class I protected plants (Table 2, Yuan et al. 2009; Zhang et al. 2015b). There are 59 key protected plants that are not protected in situ by national nature reserves, including 17 Class I protected plants.

2.4 Conservation Status of Wild Plant Species with Extremely Small Populations (PSESP) in China

According to the survey of key protected wild plant resources in the country from 1997 to 2003, the wild populations of 55 wild species such as *Abies beshanzuensis*, *Cathaya argyrophylla,* and *Pachylarnax sinica* are below the stable survival level, and they are at risk of extinction in the wild. National key protected species such as *Parrotia subaequalis* and *Yulania zenii*, as well as other economically and culturally valuable species, are also at risk of extinction (State Forestry Administration of China 2012). The State Forestry Administration has proposed the concept of Wild Plant Species with Extremely Small Populations (PSESP), which refers to the narrow or intermittent distribution of geographical distribution and the long-term disturbance of external factors to present population degradation and quantity. The number of PSESP is continuously reduced, the size of population and the number of individuals are extremely small, and are already below the minimum viable population at the stable survival limit are endangered at any time (Ren et al. 2012). On the one hand, most of these very small population species are endemic to China and have important ecological and economic values. On the other hand, the decline in the viability of wild plants, habitat destruction and loss caused by human activities, and overutilization of resources have made it extremely vulnerable to extinction (Ren et al. 2012). Given that very small populations of wild plants are among the most vulnerable biological resources, protecting them is among top priorities for biodiversity conservation. If the protection is not timely enough, the biological and genetic value of very small population species may disappear with the extinction of species before they are known and ultimately will bring immeasurable losses to ecosystems and human society. Therefore, protecting species with extremely small populations can help to delay species extinction, maintain ecological balance, preserve resources, and promote ecological sustainable development, which is of great significance to China's biodiversity conservation.

2.5 *Distribution of PSESP*

There are many species with extremely small populations in southern China, especially in the southeastern part of Yunnan, southwestern Guangxi, and southwestern Hainan Island. These regions all have more than four species. The most abundant areas are located in Lingshui and Ledong Li Autonomous County of Hainan Island, with 11 species, respectively. In Sichuan, Chongqing, Hunan, Jiangxi, Fujian, and Jilin, there are also some PSESP, with a richness of around 1.

2.6 *Assessment of the Status of Conservation of PSESP in Nature Reserves*

The NNRs have an average coverage rate of 21.5% for the distribution of PSESP, and 35 species (29% of the total) are not covered by any NNR. The average coverage rate of the PNRs to the distribution of PSESPs is 10.9%. In addition, 17 very small populations of wild plants (14% of the total) are not covered by any nature reserve.

The PSESP that are not covered by NNRs are mainly distributed in northern Yunnan, southwestern Guangxi, and Hainan Island. The PSESPs not covered by national and provincial protected areas are mainly distributed in central Sichuan, western Guizhou, and southwestern Hainan Island.

2.7 *Problems and Recommendations*

It is very urgent to take action to protect key protected plants that were not covered by national nature reserves as soon as possible. According to assessment results, key protected plants not protected by national nature reserves are mainly distributed in 50 counties and cities in 11 provinces and autonomous regions. The necessity of establishing protected areas in these areas can be considered in combination with the distribution of provincial- and county-level protected areas. At the same time, in reference to the IUCN on the classification of nature reserves, the establishment of different types of nature reserves is possible. The protection of very small populations needs to be strengthened in two ways: strengthening in situ conservation work and improving ex situ conservation. Both Yunnan and Hainan provinces have the many PSESPs which are not covered by protected areas, and threatened plants include orchids (Zhang et al. 2015a, b), so it is urgent to establish new nature reserves or give priority to upgrade existing provincial protected areas that are habitat of PSESPs and improve the management of protected areas and the effectiveness of existing protected areas.

Relying on the existing botanical gardens and breeding grounds specializing in ex situ conservation for PSESPs will not provide sufficient ex situ conservation resources. It is recommended to establish more botanical gardens in Hainan Island, Yunnan, Hunan, Sichuan, and other places, especially key counties such as Ledong and Lingshui. These together will protect the system and strengthen the function of plant conservation for very small populations.

Improve the accuracy and completeness of the original data. There are many studies of distribution of key protected plants in China; some of them are in-depth (Zhang and Ma 2008). However, the scope and targets of the first national survey of wild plant resources carried out by the State Forestry Administration were slightly insufficient. The determination of the detailed distribution location of many key protected plants relies mainly on literature and specimen data. It is recommended to further strengthen the investigation of key protected plants national wide and at the same time strengthen the investigation and accumulation of protected area background data.

Assess the effectiveness of in situ conservation of key protected plants. The current status of key protected plants is evaluated according to their natural distribution and protected status, and the in situ conservation effectiveness indicators are quantified, focusing on the research on species that are not effectively protected and improving the efficiency of in situ conservation of key protected plants.

3 Ex Situ Conservation

3.1 Overview of Ex Situ Conservation

Botanical gardens and arboreta play a key role in ex situ conservation and plant resource collection and storage. Botanical gardens and arboretum are institutions holding documented collections of living plants for the purposes of scientific research, conservation, display, and education (Wyse Jackson and Sutherland 2013; Ren and Duan 2017). Botanical gardens and arboretum conserve plant diversity ex situ and can prevent extinction through integrated conservation actions. There were over 3269 botanical collections in 180 countries around the world. Botanical gardens collected at least 105,634 species, equating to 30% of all plant species diversity, and ex situ conserved over 41% of known threatened species. However, botanical gardens in the Northern Hemisphere held 93% of species and about 76% of species absent from living collections in tropical area. Over 50% of vascular genera, but barely 5% of non-vascular genera, were conserved ex situ. Only

10% of botanical gardens conserved threatened species. Therefore, botanical gardens played a fundamental role in plant conservation, but should take actions to enhance future conservation of biodiversity (Mounce et al. 2017).

3.2 *Ex Situ Conservation in Chinese Botanical Gardens*

There were 162 botanical gardens or arboreta with an area of 102,007 ha in China. Among them, 62 botanical gardens or arboreta were located in temperate regions, 68 in subtropical regions, and 32 in tropical regions. No botanical gardens had been established in the cold and temperate regions and the Qinghai-Tibet Plateau until 2018. These botanical gardens or arboreta had established 1195 section gardens for living collection, with an area of about 5400 ha (Fig. 2). These botanical gardens also had constructed nurseries with an area of 1015 ha, plant tissue culture facilities with an area of 36,745 m^2, and seed bank or seed specimen bank with an area of 11,962 m^2 (Huang 2018).

Chinese botanical gardens or arboreta played an active role in the protection of native plant diversity (Huang 2018). The accession number for living collection and ex situ cultivated plants was 387,749 at Chinese botanical gardens or arboreta. There were 25,029 species and 316,316 taxa. Totally, 22,104 species, belonging to 2911 genera and 288 families of Chinese native plants, are ex situ cultivated at botanical gardens or arboreta. This is equivalent to 65%, 86%, and 91% of all Chinese native plant species, genera, and families, respectively (Table 3).

Fig. 2 The cycas garden at Shenzhen Fairylake Botanical Garden (Photo provided by Dr. Nan Li)

Table 3 The main families and their species in ex situ cultivation in China (Huang 2018)

Family name	Species of the family
Orchidaceae	695
Fabaceae	493
Rosaceae	566
Gesneriaceae	344
Lauraceae	255
Poaceae	257
Lamiaceae	245
Asteraceae	223
Rubiaceae	297
Gramineae	602
Dryopteridaceae	203
Ericaceae	194
Theaceae	181
Euphorbiaceae	348
Begoniaceae	176
Magnoliaceae	172
Primulaceae	169
Zingiberaceae	365
Acanthaceae	150
Liliaceae	517
Fagaceae	140
Polypodiaceae	140
Apocynaceae	137
Urticaceae	136
Moraceae	132
Sapindaceae	119
Malvaceae	118
Ranunculaceae	116
Berberiaceae	109
Caprifoliaceae	109
Rutaceae	106
Pinaceae	105
Athyriaceae	104
Cyperaceae	99

The number of endangered and threatened plants in ex situ conservation was around 1500, which was about 39% of the endangered and threatened species recorded in China (Table 4). Up to 2015, the Chinese botanical gardens or arboreta had successfully cultivated around 270 species from the *China National Key Protected Wild Plants List* (equivalent to 85% of the species on the list).

Table 4 The number of species and threatened and endangered plants under ex situ conservation in China (Huang and Zhang 2012; Huang 2018)

Botanical gardens	The number of species	The number of threatened and endangered plants
South China Botanical Garden	17,000	599
Xishuangbanna Tropical Botanical Garden	12,000	575
Kunming Botanical Garden	7000	456
Wuhan Botanical Garden	11,700	431
Nanjing Botanical Garden	7000	364
Guilin Botanical Garden	5100	313
Shenzhen Botanical Garden	12,000	278
Lushan Botanical Garden	5400	173
Beijing Botanical Garden	10,000	125
Xiamen Botanical Garden	6300	112
Total unduplicated number	–	1423

3.3 Agricultural Germplasm Resources Preservation

At present, the total amount of germplasm resources of crops preserved in China takes the second highest position in the world. The Chinese Germplasm Bank of Wild Species in Southwest China includes a seed bank, a plant in vitro bank, a DNA bank, a microbial bank, and an animal resource bank, as well as experimental research platform on plant genomics and seed biology. A total of 10,048 species of wild plant seeds were preserved in 2018 (Ren et al. 2019). The Ministry of Agriculture and Rural Affairs launched the third national survey and collection of crop genetic resources in July 2015. The aim of the campaign is to conduct a comprehensive survey of all agricultural counties, districts, and municipalities in the country, collect 100,000 germplasm resources, and safely preserve 70,000 copies. At the same time, the collection of traditional knowledge related to germplasm resources was carried out. The Ministry of Agriculture and Rural Affairs established the National Crop Germplasm Resources Sharing Service Platform at the Institute of Crop Science in Chinese Academy of Agricultural Sciences (ICS-CAAS). By the end of 2016, China had established one long-term, one duplication, and ten mid-term gene banks in which more than 785 germplasm resources of crops such as grain, vegetables, fruit trees, grass, tropical crops, fiber, oil, sugar, tea, tobacco, mulberry, green manure, and so on, totaling 481,617 collections, were safely preserved. Meanwhile, 43 field gene banks for vegetative crops were also constructed and preserved 66,385 accessions of genetic resources. In order to protect some wild relatives of crops from distinction, China has constructed 199 in situ conservation sites for more than 56 species of wild relatives of crops. In

2011, China completed the survey of medicinal plant resources and ex situ conservation and carried out the fourth national survey of traditional Chinese medicine resources. By 2018, more than 230,000 samples of medicinal materials, wax leaves, and germplasm resources had been collected. Moreover, more than 50,000 accessions of germplasm resources of trees were also collected and preserved in gene banks and field gene banks specifically for trees (Ren et al. 2019).

3.4 Problems and Recommendations

Different regional native flora are represented across China, as most botanic gardens have emphasized the collection and conservation of local or regional floras. Priority should be given to developing genetically representative collections (considering population sizes, distribution, and ecological traits) of the most critically threatened species. It is an essential problem of ex situ conservation that the number of plants planted in botanical gardens is small. Besides, many, if not most, of the collections lack sufficient genetic diversity (Ren and Duan 2017). Furthermore, 37% of the plants are cultivated only in one botanical garden, 40% in two botanical gardens, and 23% in three or more botanical gardens (Huang 2018). The program Full Coverage Conservation Plans for Native Plants in China was launched by the Chinese Union of Botanical Gardens in 2013, which also aims to safeguard national plant diversity. It will be dependent on better characterization of existing collections and long-term monitoring, to ensure the adequate genetic representation of ex situ collections.

In the future, further definition of priority taxa is needed, such as narrow endemics, sub-specific taxa, critically endangered species, taxa with known or potential future use, and wild relatives of useful plants. It should be a top priority for ex situ collections and reintroduction programs to build and manage appropriate genetic diversity in the future, besides, to strengthen the linkages between in situ and ex situ conservations at species and ecosystem levels (Fig. 3), as well as stakeholder and policy levels, and to improve national coordination of ex situ collection policies and curatorial efforts to secure research and conservation value.

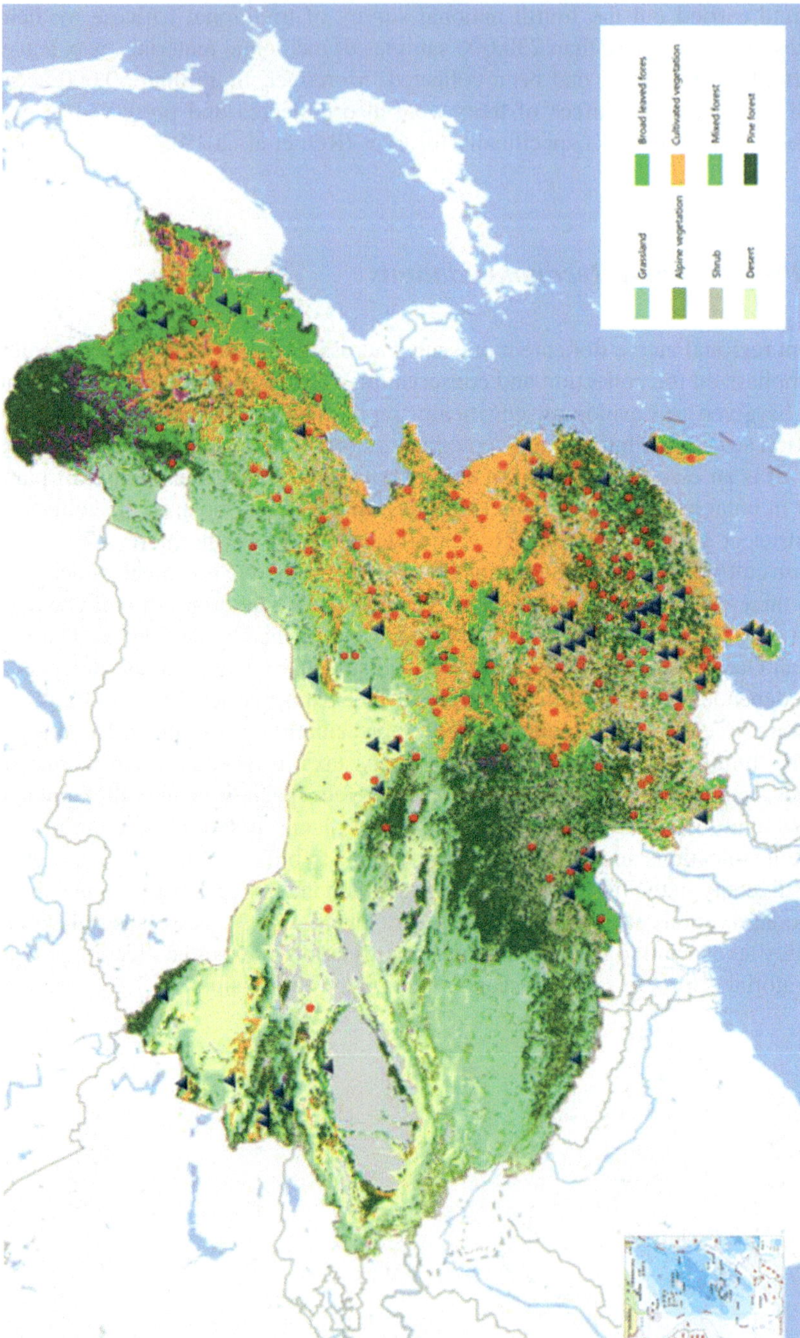

Fig. 3 Botanical reserves, botanical gardens, and vegetation distribution in China (Ren et al. 2019)

References

H.W. Huang, Plant diversity and conservation in China: Planning a strategic bioresource for a sustainable future. Bot. J. Linnean Soc. **166**, 282–300 (2011)

H.W. Huang, *The Chinese Botanical Gardens* (Chinese Forestry Press, Beijing, 2018)

H.W. Huang, Z. Zhang, Current status and prospects of *ex situ* cultivation and conservation of plants in China. Biodivers. Sci. **20**, 559–571 (2012)

J.S. Miller, H.A. Porter-Morgan, H. Stevens, et al., Addressing target two of the global strategy for plant conservation by rapidly identifying plants at risk. Biodivers. Conserv. **21**, 1877–1887 (2012)

R. Mounce, P. Smith, S. Brockington, Ex-situ conservation of plant diversity in the world's botanic gardens. Nature Plants. **3**, 795–802 (2017)

H.N. Qin, N.L. Zhao, Evaluating the threat status of higher plants in China. Biodivers. Sci. **25**(7), 689–695 (2017)

H.N. Qin, Y. Yang, S.Y. Dong, et al., List of threatened species of higher plants in China. Biodivers. Sci. **25**(7), 696–744 (2017)

H. Ren, Z.R. Duan, *The Theory and Practice of Building Scientific Botanical Gardens*, 2nd edn. (Science Press, Beijing, 2017)

H. Ren, Q.M. Zhang, H.F. Lu, H.X. Liu, Q.F. Guo, J. Wang, S.G. Jian, H.O. Bao, Wild plant species with extremely small populations require conservation and reintroduction in China. AMBIO **41**(8), 913–917 (2012)

H. Ren, S.G. Jian, H.X. Liu, Q.M. Zhang, H.F. Lu, Advances in the reintroduction of rare and endangered wild plant species. Sci. China Life Sci. **57**, 603–609 (2014)

H. Ren, H.N. Qin, Z.Y. Ouyang, et al., Progress of implementation on the global strategy for plant conservation in China (2011–2020). Biol. Conserv. **230**, 169–178 (2019)

State Forestry Administration of China, The saving and conservation program on extremely small populations in China (2012)

P. Wyse Jackson, L.A. Sutherland, Role of botanic gardens, in *Encyclopedia of Biodiversity*, ed. by S. A. Levin, vol. 6, 2nd edn., (Academic Press, Waltham, 2013)

Y.F. Yu, The national key protected wild plants in China. Plant J. **151**(5), 3–11 (1999). (in Chinese)

H. Yuan, Y.B. Zhang, H.N. Qin, et al., Current situation of in situ conservation of national key protected wild plants in China. Biodivers. Sci. **17**(3), 280–287 (2009)

Y.B. Zhang, K.P. Ma, Geographical distribution characteristics of national key protected wild plants in China. J. Appl. Ecol. **8**, 1670–1675 (2008)

Z. Zhang, Y. Yan, Y. Tian, J. Li, J.S. He, Z.R. Tang, Distribution and conservation of orchid species richness in China. Biol. Conserv. **181**, 64–72 (2015a)

Z. Zhang, J.S. He, J. Li, et al., Distribution and conservation of threatened plants in China. Biol. Conserv. **192**, 454–460 (2015b)

Reintroduction of Rare and Endangered Plants in China

Hai Ren, Hongfang Lu, Hongxiao Liu, and Zhanhui Xu

Abstract The reintroduction of species is the deliberate establishment of individuals of a species in an area and/or habitat where it has become extinct or nearly extinct. There are several procedures and guidelines for reintroductions. Reintroduction procedures for rare and endangered plant focused on "why, when, where, what, who, and how" and involves in pre-project activities, preparation and release stages, and post-release activities. Over 890 papers related to species reintroduction were published from 1980 to 2009. Worldwide, it has been attempted to reintroduce for about 700 taxa from the United States, Australia, Europe, and South Africa, with 301 plant species. About 2044 papers on plant reintroduction were searched out in the ISI database in January 2020. Most of the papers reported reintroduction cases. Nowadays, reintroduction biology has been established as an important tool for biodiversity conservation. Some papers focused on reintroduction and ecological restoration under global climate. A protocol has been established for reintroducing rare and endangered plants. There were 300 plant reintroduction projects in China by the end of 2019. These reintroductions involved 206 species, 112 of which are Chinese endemics. In addition, problems and recommendations around the reintroduction of rare and endangered plants in China are discussed.

Keywords Reintroduction · Reintroduction biology · Reintroduction procedure · Rare and endangered plant · Climate change

1 Introduction

The combination of the complementary approaches of in situ and ex situ conservation strategies to the protection and management of biological diversity has been termed as integrated conservation (Falk et al. 1996; Ren et al. 2010). Integrated

H. Ren (✉) · H. Lu · H. Liu · Z. Xu
South China Botanical Garden, Chinese Academy of Sciences, Guangzhou, China
e-mail: renhai@scbg.ac.cn; luhf@scib.ac.cn; liuhx@scbg.ac.cn; zhxu@scbg.ac.cn

H. Ren (ed.), *Conservation and Reintroduction of Rare and Endangered Plants in China*, https://doi.org/10.1007/978-981-15-5301-1_4

conservation strategies seek to (1) combine research on individual species with the scientific management of communities and ecosystems; (2) capture and maintain the genetic variation present in populations, including distinct ecotypic variants; and (3) draw together organization using different approaches and methods for the conservation of diversity (Akeroyd and Wyse 1995). Approximately 65% of vascular plant communities has been protected at natural reserves and national park systems in China using in situ conservation techniques (Huang and Zhang 2012). Besides, about 60% of plant species has been preserved in botanical gardens and other ex situ conservation facilities (Ren and Duan 2017). An introduction priority was given to endangered plant species by international convention, besides species with important economic, cultural, and ecological values (Ye et al. 2002; Ye and Chen 2004; Ren et al. 2016a, b). Together with sustainable utilization, reintroduction is not only the final goal of ex situ conservation but also the bridge between in situ and ex situ conservation (Ren et al. 2014).

Species reintroduction is the deliberate establishment of individuals of a species in an area and/or habitat where the species has become extinct or nearly extinct. The aim is to directly support the protection of the target species and to establish a viable and self-sustaining population of the species in the wild which has sufficient genetic diversity to adapt with environmental change (Maunder 1992; Ren et al. 2014; Commander et al. 2018). The objectives of a reintroduction include (1) enhancement of the long-term survival of a species; (2) reestablishment of a keystone species in an ecosystem; (3) maintenance and/or restoration of natural biodiversity; and (4) promotion of long-term economic benefits to the local and/or national economy and conservation awareness (IUCN 1998).

There are three fundamental concepts for reintroduction, i.e., augmentation, restitution, and translocation. In augmentation (referred to as reinforcement, or enhancement, or supplementation), individuals are added to an existing population, to increase the size and/or the genetic diversity of the population. In restitution (referred to as reestablishment, or restoration), individuals are introduced into an area where the species has formerly inhabited. In translocation (referred to as species introduction, assisted migration, and assisted colonization), individuals of a species are translocated from the existing distribution area to new locations (usually in the wild) that is not part of the historic distribution (Griffith et al. 1989; Zhou and Gao 2011; Ren et al. 2014). A spectrum of conservation translocation was defined by Seddon (2010), including reintroduction, reinforcement, ecological replacement, assisted colonization, and community construction. The first two concepts are focus on restoring or augmenting the population of particular species at sites within their original historic distributions. The other three concepts are focus on introducing species into areas which are not part of their original historic distributions. In addition, *resurrect* referred to the reintroduction to wild habitat of a species that has been extinct in the wild (regionally or globally) but was ex situ conserved (Seddon et al. 2014; Ren et al. 2019).

Reintroduction procedures for rare and endangered plant focused on the "six Ws": why, when, where, what, who, and how. Procedures and guidelines for reintroductions have been developed by many international organizations and scientists, e.g., the Council of Europe (1985), Botanic Gardens Conservation International (Akeroyd and Wyse 1995), the International Union for Conservation of Nature (IUCN 1998, 2009), Falk et al. (1996), Maschinski and Haskins (2012), Commander et al. (2018), Heywood et al. (2018), and Ren et al. (2019). Besides the objectives of reintroduction, these guidelines also describe the plant species suitable for reintroduction; the habitat and plant material requirements for reintroduction; the management and monitoring of the reintroduced population and species; and finally, the procedures and evaluation criteria of the reintroduction (Ren et al. 2014).

The procedures included the following points (Council of Europe 1985; Akeroyd and Wyse 1995; Falk et al. 1996; IUCN 1998, 2009; Maschinski and Haskins 2012; Li et al. 2016; Ren et al. 2014, 2019; Commander et al. 2018):

1.1 Pre-project Activities

The following activities should be undertaken prior to initiating the project:

- Document the species' status, distribution, and ecobiological research. Is reintroduction really necessary for the survival of the species?
- Ascertain threats and, when possible, take actions to remove, control, or manage them.
- If cannot justify a reintroduction, use other conservation options.
- Reintroduction cannot harm the recipient community or to existing wild populations; otherwise, consider alternative conservation strategies.
- Determine that the reintroduction is feasible legally, logistically, and socially.
- Develop a reintroduction plan and seek peer review.
- Evaluate the economics of reintroduction, and secure adequate funding to support the project.
- Construction of a multidisciplinary team with access to expert technical advice for all phases of the program.

1.2 Preparation and Release Stages

The following factors are critical to the implementation of the reintroduction project:

- Choice of release site and type.
- Evaluation of reintroduction site.
- Collection of appropriate source material (ascertain whether genetic studies are needed before conducting the reintroduction).
- Confirm that the species can be successfully propagated and that an adequate amount of high-quality, healthy, genetically diverse source material is available (at least 50 plants for a reintroduction; the theoretical minimum number of 10 seeds from each of 50 individual plants).
- Release of captive stock.

- Use good horticultural practice; seek or develop growing conditions with the intention of improving germination, establishment, and survival of next-generation seedlings.
- Good labeling and documentation is of special importance in reintroduction projects.
- Development of conservation education, community participation, and support.

1.3 Post-release Activities

The following are key post-release activities to ensure success of the program and gather information for future activities:

- Post-release monitoring is required of all or a sample of individuals.
- Study of processes of long-term adaptation by individuals and the population; collection and investigation of mortalities.
- Interventions (e.g., horticultural aid) when necessary.
- Decisions for revision, rescheduling, or discontinuation of program where necessary.
- Habitat protection or restoration to continue where necessary.
- Evaluation of cost-effectiveness and success of reintroduction techniques.
- Regular publications in scientific and popular literature.
- Finish the final report of the project.

A protocol has been established in China for the reintroduction of rare and endangered plants, involving the selection of an appropriate target plant species; the conduction of basic researches on species breeding and other biological and ecological aspects; the reintroduction of species to the wild companied with economic output; the enhancement of public awareness for conservation; and the promotion of the reintroduction projects all over the nation. As a result of this protocol, the commercialized development and conservation of several rare and endangered plant species has been realized, characterized with positive ecological, economic, and social benefits (Ren et al. 2012a).

Heywood et al. (2018) reported the lessons from species conservation and reintroduction practice as following: (1) Species conservation and reintroduction should be carried out in a well-planned systemic manner to avoid wasting efforts. (2) A variety of approaches can be launched in saving species from extinction. (3) A strategic planning for species conservation and reintroduction is lacking at both global and national/subnational levels. (4) Natural reserves is taken as the primary strategy by many countries for in situ conservation of threatened species, with no further targeted action to remove threats away from these species. (5) It is essential to distinguish the presence of a species in a protected area from its long-term persistence there as a viable population(s). (6) Most of the threatened species occur outside reserve areas, although most known as recovery plans are for species which are found in reserved areas. (7) For endangered species within and outside reserved areas, species-specific actions are needed to secure the long-term standing of viable populations. (8) Conservation of target species outside reserved areas can be undertaken in many ways. (9) Species reintroduction is urgently needed in tropical nationals. (10) Species conservation and reintroduction is a specifically cooperative process, essentially involving a wide range of participants and disciplines. (11) Community conservation actions are important in protecting endangered species, which deserves higher recognition. (12) Clarification of the threats to a species is an essential requirement in developing reintroduction strategies for endangered species. (13) Reintroduction objectives and evaluation standards need to be clarified as a critical part of a reintroduction plan. (14) Monitoring at all stages is another key component of recovery planning and action, which must be continued as a company of the aftercare.

Armstrong and Seddon (2008) proposed ten key questions for reintroduction biology, with different questions focusing at the population establishment and persistence, metapopulation, and ecosystem level: (1) How is establishment probability affected by the size and composition of the release group? (2) How are post-release survival and dispersal affected by pre- and post-release management? (3) What habitat conditions are needed for persistence of the reintroduced population? (4) How will genetic makeup affect persistence of the reintroduced population? (5) How heavily should source populations be harvested? (6) What is the optimal allocation of translocated individuals among sites? (7) Should translocation be used to compensate for isolation? (8) Are the target species/taxon and its parasites native to the ecosystem? (9) How will the ecosystem be affected by the target species and its parasites? (10) How does the order of reintroductions affect the ultimate species composition? Ames et al. (2020) suggest that the functional composition of the biotic community should be incorporated into site selection to improve reintroduction success.

The discipline of species recovery is currently considered as independent of conservation biology. The advancement of species reintroduction was highlighted by many recent publications, e.g., *Restoring Diversity: Strategies for the Reintroduction of Endangered Plants* (Falk et al. 1996); *Reintroduction Biology: Integrating Science and Management* (Ewen et al. 2012); and *Plant Reintroduction in a Changing Climate: Promises and Perils* (Maschinski and Haskins 2012).

Reintroduction biology is a field of scientific research that aims to inform translocations of endangered species (Taylor et al. 2017). Reintroduction research increasingly addresses a priori hypotheses but remains largely focused on short-term population establishment. Similarly, studies that directly assist decisions by explicitly comparing alternative management actions remain a minority. A small set of case studies demonstrate full integration of research in the reintroduction decision process. Conservationists encourage the use of tools that embed research indecision-making, particularly the explicit consideration of multiple management alternatives because this is the crux of any management decisions.

2 Reintroduction of Rare and Endangered Plants Around the World

The number of reintroduction-related papers in peer-reviewed journals increased from <5 per year during the early 1990s to >50 per year currently. Polak and Saltz (2011) identified 890 papers related with species reintroduction based on a searching in the ISI database from 1980 to 2009. Worldwide, reintroductions have been undertaken for around 700 taxa from the United States, Australia, Europe, and South Africa, with 301 plant species (Maschinski and Haskins 2012). About 2044 papers on plant reintroduction were searched out in the ISI database in January 2020.

Maschinski and Haskins (2012) evaluated the 301 reintroductions cases for 128 plant taxa by considering the survival of propagule, persistence of population, and the recruitment of next generation. Approximately 20% of the projects involved reintroduction, with 30% involving augmentation. They also found that endemic species had higher probability to be reintroduced successfully than widespread species, while other factors did not have significant impact. However, this may be due to higher investments associated with the conservation of endemic species.

Godefroid et al. (2011) carried out a global review on plant reintroduction. They collected a total of 249 cases involving 172 taxa. They found that the average survival rate was more than 52%, and the flowered or fruited rate was less than 20%. They also indicated that demographic, genetic, and ecological factors influenced success rates in plant reintroductions and present valuable recommendations to improve plant reintroduction success.

In the United States, Albrecht et al. (2011) analyzed 49 cases of plant reintroductions at the CPC International Reintroduction Registry (CPCIRR), and found a 92% survival rate of the populations, with 76% of them reached reproductive maturity, 33% of them produced a next generation, and 16% of them produced a next generation with reproductive individuals. It is showed by the CPCIRR database that some reintroduced populations have stood there for over 24 years. They also explored that a more than 4-year data is needed to document population sustainability, especially for long-lived perennials.

In Australia, Silcock (2018) identified more than 50 plant species that have been deliberately translocated by Aboriginal Australians, spanning a range of life forms and much of the continent. The vast majority were important food species, while others were valued for plant materials. Over one-third had ceremonial or cultural importance, with translocations of these often occurring as part of specific ceremonies (Commander et al. 2018). Sheean et al. (2012) assessed 54 reintroductions of plant species, the majority of which were reintroductions (52%) and supplementations (30%). They found 25 successful versus 14 failed cases, where the unsuccessful cases occurred due to ineffective predator control.

Constraints to plant reintroduction around the world include limited sectorial collaboration and coordination, lack of mainstreaming, a lack of information, limited institutional integration, lack of tools and technologies, and limited financial and human resources. In order to strengthen plant reintroduction around the world, the world plant diversity should be well understood, documented, and recognized; plant diversity should be urgently and effectively conserved (in situ, ex situ, and reintroduction) and sustainably used; education and awareness about plant diversity and its role in securing the sustainable development and the welfare of all life on earth should be promoted; and the capacities and public participation should be developed. In the future, all countries may formulate an effective and participatory national plant conservation strategy or action plan, to invest more to plant conservation and reintroduction; to develop online information systems on their floras, natural habitats, reintroduction, and knowledge sharing; to develop science base and technologies on plant conservation and reintroduction; to assess plant conservation status; to integratively conserve wild plants with in situ conservation, ex situ conservation, and reintroduction together; to effectively protect each ecological region; to decrease anthropogenic pressures on ecosystems and species and control invasive species; and to improve capability on plant conservation and reintroduction.

3 Reintroduction of Rare and Endangered Plants in China

The concept of reintroduction as a tool for species conservation has gained popularity among Chinese practitioners and the public in general, as evidenced by the frequent media reports since 2008 (Liu et al. 2004, 2012). The Conservation Program for Wild Plants Species with Extremely Small Populations (2011–2015), initiated by the State Forestry Administration of China in 2012, signals acceptance of reintroduction as an integral tool for species recovery in China (Ren et al. 2012b; Liu et al. 2015).

The State Forestry Administration of China formulated the Conservation Program for Wild Plants Species with Extremely Small Populations in China (2010–2015) as their 2012–2015 operational plan in 2012. The program tried to conserve 120 wild plants speices with extremely small populations. The concept of the wild plants species with extremely small populations was based on recognition of species with (1) unique topography and habitat; (2) extremely limited distribution range; (3) enormous pressure due to habitat destruction; (4) habitat in plant diversity hotspot areas with poor economic development; and (5) the importance of culture and traditional Chinese values toward nature (Ren et al. 2012a; Ma et al. 2013).

The program included in situ conservation (individuals' location, numbering, labeling, document management, monitoring, management and nursing, habitat restoration, infrastructure establishment, etc.), near situ conservation (population management and monitoring, infrastructure construction, etc.), ex situ conservation (nursery construction for seedling propagation, propagation experiments, population establishment, document management, etc.), germplasm gene preservation (seed protection, collection management, seed and pedigree management), reintroduction (species selection, site selection, population management, monitoring, etc.), and capacity building (propaganda and education, personnel training, scientific research, cooperation, and exchange). By 2015, the program had invested 124 million CNY (approximately $17.9 million USD) and had carried out field survey on 90 species of 120 species wild plants in China. A total of 51 species had been effectively protected in situ, 20 species had been successfully protected in situ, 56 species had been successfully reproduced to obtain seedlings, and 26 species had been reintroduced to the wild.

There were 300 plant reintroduction projects in China by the end of 2019 (Table 1, Fig. 1). These reintroductions involved 206 species, 112 of which are Chinese endemics (Table 1). There were 166 species of angiosperms, 34 gymnosperms, and 2 Pteridophyta, representing a total of 60 families and 136 genera. The Orchidaceae has the largest number of translocated species (43 spp.), followed by Magnoliaceae (21 spp., Fig. 2), Pinaceae (12 spp.), Oleaceae (9 spp.), Lauraceae (7 spp.), Aceraceae (7 spp.), Fabaceae (6 spp.), Betulaceae (5 spp.), Leguminosae (5 spp.), Podocarpaceae (5 spp.), Rutaceae (5 spp.), and Theaceae (5 spp.). Among them, 60 species were herbaceous plants, 24 were shrubs (including one palm and two cycads), 2 were vines, and the rest (121) were trees (Fig. 3).

Problems and difficulties around the reintroduction of rare and endangered plants in China are as following:

1. China's existing laws and regulations on wild plant conservation need to be further improved. For example, the "list of national key protected wild plants" needs to be updated, and many wild plants with extremely small population fail to be "legally" included in the scope of national key protection. Due to lax law enforcement, the phenomenon of destroying wild plants often occurs. For wild plants with extremely small population distributed in nature reserves, in order to ensure their survival and reproduction, and expand their population size, appropriate artificial intervention is needed to help flowering, fruiting, seedling emergence, planting, etc., as well as habitats management. Due to the limitations of the regulations on nature reserves, there is an awkward state that they cannot be effectively protected.
2. Insufficient investment in plant conservation. There are many kinds of rare and endangered plants in China, and most of them are distributed in remote mountains. Although the central government and local governments have increased the capital investment in the protection of rare and endangered plants, compared with the actual demand, the capital investment is far from enough.
3. Scientific research is not systematic enough. At present, flora of China has been published, and the endangered status of Chinese plants has been assessed, and the distribution of wild plants in China has been roughly figured out. However, the systematic scientific research on these rare and endangered plants is not enough, and the ecological and biological characteristics of many species are not disclosed. In addition, there is a lack of technical standards or guidelines for reintroduction.
4. The awareness and ability of society to protect is not enough. At present, the governments at all levels do not know the importance of wild plants with extremely small population in place, the citizens' awareness of plant protection needs to be improved, and the professional ability of the managers of wild plant protection needs to be strengthened too.

Table 1 Number of conservation translocation projects in China

No.	Species	Chinese name	Endangered sp. listing (IUCN or China)	Genus	Family	Plant life form	Natural distribution (mainly based on Flora of China)	Chinese endemic	Endemic Genus	Endemic Family	Monogenous species	Tropical or subtropical species	Climate types of habitats	Ecosystem types of habitats
1	*Abies beshanzuensis*	百山祖冷杉	China RL (2013)-CR, China TSLHP (2017)[a]-CR, IUCN(2014)-CR	*Abies*	Pinaceae	Tree	Baishan National Nature Reserve, Qingyuan County, Zhejiang Province	Yes	No	No	No	Yes	Subtropical monsoon climate	Forest
2	*Acanthephippium sylhetense* Lindl.	坛花兰	China RL (2013)-VU, China TSLHP (2017)[a]-VU	*Acanthephippium*	Orchidaceae	Herb	Southern Yunnan (Mengla, Jinghong) and Taiwan (Taipei, Wuli, Daxueshan, Kaohsiung, Lanyu)	No	No	No	No	Yes	Tropical monsoon	Under dense forest or ravine
3	*Acer hookeri* Miq.	锐齿槭	None	*Acer*	Aceraceae	Tree	Southern Tibet	No	No	No	No	No	Plateau mountain climate	Forest
4	*Acer miaotiense*	庙台槭	None	*Acer*	Aceraceae	Tree	Southwest Shaanxi and Southeast Gansu	No	No	No	No	No	Temperate monsoon climate	Forest
5	*Acer yangbiense*	漾濞槭	China RL (2013)-CR, China TSLHP (2017)[a]-CR, Acer RL-CR	*Acer*	Aceraceae	Tree	Yunnan Province (Cangshan)	Yes	No	No	No	Yes	Subtropical monsoon	Forest/ mountain
6	*Acer yangjuechi*	羊角槭	National key protected wild plants of China-II	*Acer*	Aceraceae	Tree	Northwest Zhejiang Province	Yes	No	No	No	Yes	Subtropical monsoon climate	Forest

(continued)

Table 1 (continued)

No.	Species	Chinese name	Endangered sp. listing (IUCN or China)	Genus	Family	Plant life form	Natural distribution (mainly based on Flora of China)	Chinese endemic	Endemic Genus	Endemic Family	Monogenous species	Tropical or subtropical species	Climate types of habitats	Ecosystem types of habitats
7	*Acrocarpus fraxinifolius*	顶果树	China RL (2013)-VU, China TSLHP (2017)[a]-VU	*Acrocarpus*	Leguminosae	Tree	Guangxi (Longlin, Tianlin, Debao) and Yunnan (Jingdong, Xishuangbanna, Hekou)	No	No	No	No	Yes	Subtropical monsoon	Sparse forest
8	*Adiantum reniforme* var. *sinense*	荷叶铁线蕨	National key protected wild plants of China-I, China RL (2013)-CR, IUCN Red List (2019)-LC	*Adiantum*	Adiantaceae	Herb	Sichuan (Wanxian County, Fuling County, Shizhu County)	Yes	No	No	No	Yes	Subtropical monsoon	Forest
9	*Aesculus wilsonii*	天师栗	None	*Aesculus*	Hippocastanaceae	Tree	Southwest Henan, West Hubei, West Hunan, West Jiangxi, North Guangdong, Sichuan, Guizhou, and Northeast Yunnan	Yes	No	No	No	No	Subtropical monsoon	Forest
10	*Amentotaxus argotaenia* (Hance) Pilger	穗花杉	China TSLHP (2017)[a]-VU, IUCN (2014)-NT	*Amentotaxus*	Taxaceae	Shrub	Northwest of Jiangxi, West and southwest of Hubei, Southeast and central of Hunan, Sichuan, Southeast of Tibet, south of Gansu, Guangxi, Guangdong	Yes	No	No	No	Yes	Subtropical monsoon and temperate monsoon climate	Forest

11	*Amesiodendron tienlinense* H. S. Lo	田林细子龙	None	*Amesiodendron*	Sapindaceae	Tree	South Guizhou and Northwest Guangxi	Yes	No	No	No	Yes	Subtropical monsoon	Dense forest or sparse forest
12	*Asteropyrum peltatum*	星果草	China RL (2013)-VU, China TSLHP (2017)[a]-VU	*Asteropyrum*	Ranunculaceae	Herb	Northwest Yunnan, Sichuan, and West Hubei	Yes	Yes	No	No	Yes	Plateau mountain climate	Forest
13	*Berberis delavayi* Schneid.	雪山小檗	None	*Berberis*	Berberidaceae	Tree	Null	No	No	No	No	No	Subtropical monsoon	Forest
14	*Bretschneidera sinensis* Hemsl.	伯乐树	National key protected wild plants of China-I, China RL (2013)-NT, IUCN (2014)-EN	*Bretschneidera*	Bretschneideraceae	Tree	Scattered in provinces and regions south of the Yangtze River	Yes	Yes	Yes	Yes	Yes	Subtropical marine wind climate	Forest
15	*Burretiodendron esquirolii* (Levl.) Rehd.	柄翅果	National key protected wild plants of China-II, China RL (2013)-VU, IUCN (2014)-VU, China TSLHP (2017)[a]-VU	*Burretiodendron*	Tiliaceae	Tree	Southeast, Luodian, Guizhou, Ceheng to Hongshuihe, Guangxi	Yes	No	No	No	Yes	Subtropical monsoon	Forest
16	*Caesalpinia sappan*	苏木	National key protected wild plants of China-II, IUCN (2014)-LR/lc	*Caesalpinia*	Leguminosae	Tree	Jinsha River Valley (Yuanmou, Qiaojia) and Honghe River Valley in Yunnan Province	No	No	No	No	Yes	Subtropical monsoon	Forest

(continued)

Table 1 (continued)

No.	Species	Chinese name	Endangered sp. listing (IUCN or China)	Genus	Family	Plant life form	Natural distribution (mainly based on Flora of China)	Chinese endemic	Endemic Genus	Endemic Family	Monogenous species	Tropical or subtropical species	Climate types of habitats	Ecosystem types of habitats
17	*Calanthe argenteostriata*	银带虾脊兰	National key protected wild plants of China-II, China RL (2013)-LC	*Calanthe*	Orchidaceae	Herb	North Guangdong (Conghua), Southwest Guangxi (Longzhou), Southwest Guizhou (Xingyi, Anlong), and Southeast Yunnan (Jinping, Malipo)	No	No	No	No	Yes	Subtropical monsoon	Forest
18	*Calanthe sieboldii Decne.*	大黄花虾脊兰	China RL (2013)-CR, China TSLHP (2017)[a]-CR	*Calanthe*	Olacaceae	Herb	North Taiwan (Taipei, Hsinchu, etc.) and Southwest Hunan (Xinning)	No	No	No	No	Yes	Tropical rainforest climate	Forest
19	*Calycanthus chinensis*	夏蜡梅	National key protected wild plants of China-II, China RL (2013)-EN, China TSLHP (2017)[a]-EN, IUCN Plant (1997)-V	*Calycanthus*	Calycanthaceae	Shrub	Changhua and Tiantai, Zhejiang Province	Yes	No	No	No	Yes	Subtropical monsoon	Ditch edge of mountain
20	*Calycopteris floribunda* (Roxb.) Lam.	萼翅藤	National key protected wild plants of China-I	*Calycopteris*	Combretaceae	Vine	Nabangba, Yingjiang County, Dehong Prefecture, Yunnan Province	No	No	No	No	Yes	Subtropical monsoon and tropical monsoon climate	Forest

21	*Camellia azalea*	杜鹃红山茶	China RL (2013)-CR, IUCN (2016)-CR, China T SLHP (2017)[a] -CR	*Camellia*	Theaceae	Shrub	Ehuangzhang Nature Reserve in Yangchun County, Yangjiang City, in southwest Guangdong Province	Yes	No	No	No	Yes	Subtropical monsoon	Forest in mountainous region
22	*Camellia fascicularis*	云南金花茶	China RL (2013)-CR, China TSLHP (2017)[a]-CR, IUCN (2016)-CR	*Camellia*	Theaceae	Tree	Yunnan Province, China	Yes	No	No	No	Yes	Subtropical monsoon or tropical rainforest climate	Forest
23	*Camellia grijsii*	长瓣短柱茶	None	*Camellia*	Theaceae	Shrub	Fujian, Wuxi, Lichuan, Hubei, and Northern Guangxi	Yes	No	No	No	Yes	Subtropical monsoon	Forest
24	*Carpinus putoensis*	普陀鹅耳枥	China RL (2013)-CR, China TSLHP (2017)[a]-CR, IUCN (2014)-CR	*Carpinus*	Betulaceae	Tree	West side of Huiji Temple, Fuding Mountain, Putuo Island, Zhoushan City, Zhejiang Province	Yes	No	No	No	Yes	Subtropical monsoon climate	Forest
25	*Carpinus tientaiensis*	天台鹅耳枥	China RL (2013)-CR, China TSLHP (2017)[a]-CR, IUCN (2014)-CR	*Carpinus*	Betulaceae	Tree	Taihuading National Forest Park, Zhejiang Province	Yes	No	No	No	Yes	Subtropical monsoon climate	Forest
26	*Cathaya argyrophylla*	银杉	China RL (2013)-EN, China TSLHP (2017)[a]-VU, IUCN (2014)-VU	*Cathaya*	Pinaceae	Tree	Guizhou, Chongqing, Guangxi	Yes	Yes	No	Yes	Yes	Subtropical monsoon	Forest

(continued)

Table 1 (continued)

No.	Species	Chinese name	Endangered sp. listing (IUCN or China)	Genus	Family	Plant life form	Natural distribution (mainly based on Flora of China)	Chinese endemic	Endemic Genus	Endemic Family	Monogenous species	Tropical or subtropical species	Climate types of habitats	Ecosystem types of habitats
27	*Cephalotaxus oliveri*	篦子三尖杉	China RL (2013)-VU, China TSLHP (2017)[a]-VU	*Cephalotaxus*	Cephalotaxaceae	Shrub	300–1800 m above sea level in the north of Guangdong, east of Jiangxi, north-west of Hunan, northwest of Hubei, South and west of Sichuan, Southeast and northeast of Gui-zhou and Yunnan	No	No	No	No	Yes	Subtropical monsoon	Forest
28	*Cercidiphyllum japonicum*	连香树	National key protected wild plants of China-II, China RL (2013)-LC, IUCN (2014)-LR/nt	*Cercidiphyllum*	Cercidiphyllaceae	Tree	Southwest Shanxi, Henan, Shaanxi, Gansu, Anhui, Zhejiang, Jiangxi, Hubei, and Sichuan	No	No	No	No	No	Temperate monsoon climate	Forest
29	*Changnienia amoena*	独花兰	China RL (2013)-EN, IUCN (2014)-EN, China TSLHP (2017)[a]-EN	*Changnienia*	Orchidaceae	Herb	Southern Shaanxi, Jiangsu, Anhui, Zhejiang, Jiangxi, Hubei, Hunan, and Sichuan (Wushan, Beichuan, Guangyuan, Bazhong, and Maowen)	Yes	Yes	No	Yes	Yes	Subtropical monsoon	Forest

30	*Cheirostylis chinensis*	中华叉柱兰	China RL (2013)-LC, China TSLHP (2017)[a]-LC	*Cheirostylis*	Orchidaceae	Herb	Taiwan, Hong Kong, Guangxi, Guizhou	No	No	No	No	Yes	Subtropical monsoon	Forest
31	*Chuanminshen violaceum* Sheh et Shan	川明参	China RL (2013)-EN, China TSLHP (2017)[a])-EN	*Chuanminshen*	Umbelliferae	Herb	Pingwu, Beichuan, Nanchuan, Yichang, and Dangyang, Sichuan Province	Yes	No	No	Yes	Yes	Subtropical monsoon and temperate monsoon climate	Grassland/ waterside
32	*Cibotium barometz*	金毛狗	National key protected wild plants of China-II, China RL (2013)-LC, China TSLHP (2017)[a]-EN	*Cibotium*	Dicksoniaceae	Herb	Yunnan, Guizhou, southern Sichuan, Guangdong, Guangxi, Fujian, Taiwan, Hainan Island, Zhejiang, Jiangxi, and southern Hunan (Jianghua County)	No	No	No	No	Yes	Subtropical monsoon	River valley
33	*Cinnamomum japonicum*	天竺桂	China RL (2013)-VU, China TSLHP (2017)[a]-VU, IUCN (2014)-LR/nt	*Cinnamomum*	Lauraceae	Tree	Jiangsu, Zhejiang, Anhui, Jiangxi, Fujian, and Taiwan	No	No	No	No	Yes	Subtropical monsoon	Forest
34	*Cinnamomum platyphyllum*	阔叶樟	China RL (2013)-VU, China TSLHP (2017)[a]-VU	*Cinnamomum*	Lauraceae	Tree	East Sichuan (Nanchuan, Bazhong, Chengkou)	No	No	No	No	No	Subtropical monsoon	Forest
35	*Cinnamomum septentrionale*	银木	China RL (2013)-LC	*Cinnamomum*	Lauraceae	Tree	Western Sichuan, southern Shaanxi, and southern Gansu	No	No	No	No	No	Subtropical monsoon and temperate monsoon climate	Forest

(continued)

Table 1 (continued)

No.	Species	Chinese name	Endangered sp. listing (IUCN or China)	Genus	Family	Plant life form	Natural distribution (mainly based on Flora of China)	Chinese endemic	Endemic Genus	Endemic Family	Monogenous species	Tropical or subtropical species	Climate types of habitats	Ecosystem types of habitats
36	*Cleisostoma nangongense*	南贡隔距兰	National key protected wild plants of China-II, China RL (2013)-VU, China TSLHP (2017)[a]-VU	*Cleisostoma*	Orchidaceae	Herb	Southern Yunnan (Mengla)	Yes	No	No	No	Yes	Tropical monsoon	Under dense forest or ravine
37	*Cleisostoma paniculatum* (Ker-Gawl.) Garay	大序隔距兰	China RL (2013)-LC, China TSLHP (2017)[a]-LC	*Cleisostoma*	Orchidaceae	Herb	East Jiangxi (Shangyou), Fujian (Nanjing, Xianyou, Yongtai, Chongan, Nanping, Yunxiao, Longyan), Taiwan (Taipei area), South to North Guangdong (Luofushan, Yingde, Lechang, Qujiang, Lianshan, Ruyuan), Hong Kong, Hainan (Baoting), Guangxi (Rongshui, Jinxiu, Yongfu, Shangsi), South to central	No	No	No	No	Yes	Subtropical monsoon	Forest

							Sichuan (Emeishan, Pingshan), East Guizhou Ministry (Fanjingshan), southeast to North Yunnan (Xichou, Pingbian, Maguan, Menghai, Binchuan, Yongsheng)						
38	*Cleisostoma williamsonii* (Rchb. F.) Garay	红花隔距兰	None	*Cleisostoma*	Orchidaceae	Herb	Guangdong (Gaoyao, Yunfu, Dinghushan), Hainan (Sanya City, Baoting, Lingshui, Qiongzhong, Baisha, Danxian, Ding'an), Guangxi (Lingyun, Longzhou, Rongxian, Wuming, Jingxi, etc.), Southwest Guizhou (Cexiang), Southeast Yunnan to West (Pingbian, Jinping, Funing, Jinghong, Fengqing, Menghai)	No	No	No	No	Yes	Subtropical monsoon	Forest

(continued)

Table 1 (continued)

No.	Species	Chinese name	Endangered sp. listing (IUCN or China)	Genus	Family	Plant life form	Natural distribution (mainly based on Flora of China)	Chinese endemic	Endemic Genus	Endemic Family	Monogenous species	Tropical or subtropical species	Climate types of habitats	Ecosystem types of habitats
39	*Coptis chinensis*	黄连	National key protected wild plants of China-II, IUCN Plants (1997)-V	*Coptis*	Ranunculaceae	Herb	Sichuan, Guizhou, Hunan, Hubei, southern Shaanxi	Yes	No	No	No	Yes	Subtropical monsoon and temperate monsoon climate	Forest
40	*Coptis quinquesecta*	五裂黄连	China RL (2013)-CR, China TSLHP (2017)[a]-CR	*Coptis*	Ranunculaceae	Herb	Jinping, Yunnan Province, China	Yes	No	No	No	Yes	Subtropical monsoon	Forest
41	*Corylus chinensis*	华榛	China RL (2013)-LC, IUCN Red List (2018)-LC	*Corylus*	Betulaceae	Tree	Southwest of Yunnan Province and Sichuan Province	No	No	No	No	No	Subtropical monsoon	Forest
42	*Craigia yunnanensis*	滇桐	China RL (2013)-EN, China TSLHP (2017)[a]-EN, IUCN (2014)-EN	*Craigia*	Tiliaceae	Tree	South Yunnan, South Guizhou and Southwest Guangxi	Yes	No	No	No	Yes	Subtropical monsoon	Forest
43	*Cunninghamia unicanaliculata*	德昌杉	IUCN Plants (1997)-E	*Cunninghamia*	Taxodiaceae	Tree	Dechang County, Liangshan Yi Autonomous Prefecture, Sichuan Province	Yes	No	No	No	Yes	Subtropical monsoon	Forest

44	*Cupressus chengiana*	岷江柏木	National key protected wild plants of China-II, China TSLHP (2017)[a]-VU, IUCN Red List-V	*Cupressus*	Cupressaceae	Tree	Western and Northern Sichuan (Maoxian, Wenchuan, Lixian, Dajin, and Xiaojin in the upper reaches of Minjiang River) and southern Gansu (Zhouqu, Shimen, and Wudu)	Yes	No	No	No	No	Subtropical monsoon and temperate monsoon climate	Forest
45	*Cycas debaoensis*	德保苏铁	China RL (2013)-CR, IUCN (2014)-CR, China TSLHP (2017)[a]-CR	*Cycas*	Cycadaceae	Shrub	Debao, Guangxi Province	Yes	No	No	No	Yes	Subtropical monsoon	No wild population
46	*Cycas diannanensis*	滇南苏铁	China RL (2013)-CR	*Cycas*	Cycadaceae	Tree	South of Gejiu City, Yunnan Province	Yes	No	No	No	Yes	Subtropical monsoon and tropical monsoon climate	Forest or grassland
47	*Cycas fairylakea*	仙湖苏铁	National key protected wild plants of China-I, China RL (2013)-CR, China TSLHP (2017)[a]-CR	*Cycas*	Cycadaceae	Tree	Guangxi, Guangdong, and other places	Yes	No	No	No	No	Subtropical monsoon	Forest
48	*Cycas segmentifida* D. Y. Wang and C. Y. Deng	叉孢苏铁	National key protected wild plants of China-I, China RL (2013)-EN, IUCN (2014)-VU, China TSLHP (2017)[a]-EN	*Cycas*	Cycadaceae	Shrub	Ceheng County, Wangmo County, Xilin County, Leye County, Guangxi Zhuang Autonomous Region and Funing County, Yunnan Province	Yes	No	No	No	No	Subtropical monsoon	Forest

(continued)

Table 1 (continued)

No.	Species	Chinese name	Endangered sp. listing (IUCN or China)	Genus	Family	Plant life form	Natural distribution (mainly based on Flora of China)	Chinese endemic	Endemic Genus	Endemic Family	Monogenous species	Tropical or subtropical species	Climate types of habitats	Ecosystem types of habitats
49	*Cyclobalanopsis sichourensis*	西畴青冈	China RL (2013)-CR, China TSLHP (2017)[a]-CR	*Cyclobalanopsis*	Fagaceae	Tree	Funing, Xichou, and Ceheng of Yunnan Province	Yes	No	No	No	Yes	Subtropical monsoon	Forest
50	*Cymbidium aloifolium*	纹瓣兰	China RL (2013)-NT	*Cymbidium*	Orchidaceae	Herb	Guangdong, Guangxi, Guizhou, and southeast to South Yunnan	No	No	No	No	Yes	Subtropical monsoon	Open forest/ bush
51	*Cymbidium bicolor* L. subsp. *obtusum*	硬叶兰	None	*Cymbidium*	Orchidaceae	Herb	Guangdong, Hainan, Guangxi, Guizhou, and southwest to South Yunnan	No	No	No	No	Yes	Tropical monsoon	Forest
52	*Cymbidium goeringii*	春兰	China RL (2013)-VU, China TSLHP (2017)[a]-VU	*Cymbidium*	Orchidaceae	Herb	Southern Shaanxi, southern Gansu, Jiangsu, Anhui, Zhejiang, Jiangxi, Fujian, Taiwan, Southern Henan, Hubei, Hunan, Guangdong, Guangxi, Sichuan, Guizhou, Yunnan	No	No	No	No	Yes	Subtropical monsoon and temperate monsoon climate	Forest

53	*Cymbidium mannii*	硬叶兰	China RL (2013)-NT	*Cymbidium*	Orchidaceae	Herb	Guangdong, Hainan, Guangxi, Guizhou, and southwest to South Yunnan	No	No	No	No	Yes	Subtropical monsoon or tropical rainforest climate	Forest
54	*Cymbidium tortisepalum Fukuy.* var. *longibracteatum* (Y. S. Wu et S. C. Chen) Y. S. Wu et S. C. Chen	春剑	China RL (2013)-EN, China TSLHP (2017)[a]-EN	*Cymbidium*	Orchidaceae	Herb	Sichuan, Guizhou, and Yunnan	No	No	No	No	Yes	Subtropical monsoon	Forest
55	*Cymbidium tracyanum*	西藏虎头兰	China RL (2013)-LC, IUCN (2014)-LC	*Cymbidium*	Orchidaceae	Herb	Southwest of Guizhou (Ceheng), southwest to southeast of Yunnan, and southeast of Tibet	No	No	No	No	Yes	Subtropical monsoon	Forest
56	*Cypripedium calceolus*	杓兰	China RL (2013)-NT, IUCN Red List (2019)-VU	*Cypripedium*	Orchidaceae	Herb	Heilongjiang (Yichun dailing), Eastern Jilin, Liaoning, and Northeast Inner Mongolia (Daxinganling)	No	No	No	No	Yes	Temperate monsoon climate	Forest edge or grassland
57	*Davidia involucrata* var. *involucrata*	珙桐	National key protected wild plants of China-I, China RL (2013)-LC, IUCN (2014)-LR/cd	*Davidia*	Davidiaceae	Tree	The west of Hubei Province, the west of Hunan Province, Sichuan Province, and the north of Guizhou Province and Yunnan Province	Yes	Yes	No	Yes	Yes	Evergreen broad-leaved deciduous broad-leaved mixed forest	Forest
58	*Davidia involucrata* var. *vilmoriniana*	光叶珙桐	National key protected wild plants of China-I, IUCN (2014)-VU	*Davidia*	Nyssaceae	Tree	Western Hubei, Western Hunan, Sichuan, and Northern Guizhou and Yunnan	Yes	Yes	No	Yes	Yes	Subtropical monsoon and temperate monsoon climate	Forest

(continued)

Table 1 (continued)

No.	Species	Chinese name	Endangered sp. listing (IUCN or China)	Genus	Family	Plant life form	Natural distribution (mainly based on Flora of China)	Chinese endemic	Endemic Genus	Endemic Family	Monogenous species	Tropical or subtropical species	Climate types of habitats	Ecosystem types of habitats
59	*Dendrobium aphyllum*	兜唇石斛	IUCN Red List-Least Concern/ CITES Appendix II	*Dendrobium*	Orchidaceae	Herb	Northwest of Guangxi (Longlin, Xilin, Leye), southwest of Guizhou (Xingyi), southeast to west of Yunnan (Funing, Jianshui, Jinping, Mengla, Menghai, Lushui, etc.)	No	No	No	No	Yes	Subtropical monsoon	Forest
60	*Dendrobium aurantiacum*	叠鞘石斛	IUCN Red List 1997-Extinct/ Endangered	*Dendrobium*	Orchidaceae	Herb	Hainan (Bawangling), southwest to northwest of Guangxi (Lingyun, Leye, Fengshan, Jingxi, Debao, Napo), South to southwest of Guizhou (Xingyi, Luodian, Pingtang, Anlong, Guanling, Huishui), southeast to northwest of Yunnan (Pingbian,	No	No	No	No	Yes	Subtropical monsoon	Forest

							Yanshan, Jianshui, Menghai, Fengqing, Cangyuan, Lancang, Gengma, zhenkang, Tengchong, Gongshan, Lijiang, Weixi, Deqin, etc. Land)							
61	*Dendrobium huoshanense*	霍山石斛	National key protected wild plants of China-I, IUCN (2014)-CR	*Dendrobium*	Orchidaceae	Tree	Southwest Henan (Nanzhao), Southwest Anhui (Huoshan), Jiangxi (Wuyuan)	Yes	No	No	No	Yes	Subtropical monsoon	Mountain
62	*Dendrobium lindleyi*	聚石斛	None	*Dendrobium*	Orchidaceae	Herb	East (Xinyi, Enping, Luofushan), Hong Kong, Hainan (Sanya, Lingshui, Baisha, Qiongzhong, Chengmai), Guangxi (Xilin, Daxin, Longzhou, Tianlin, Jingtian, Bobai, Yulin, Baise), South-west Guizhou (Ceheng)	No	No	No	No	Yes	Subtropical monsoon	Forest

(continued)

Table 1 (continued)

No.	Species	Chinese name	Endangered sp. listing (IUCN or China)	Genus	Family	Plant life form	Natural distribution (mainly based on Flora of China)	Chinese endemic	Endemic Genus	Endemic Family	Monogenous species	Tropical or subtropical species	Climate types of habitats	Ecosystem types of habitats
63	*Dendrobium loddigesii*	美花石斛	China RL (2013)-VU, China TSLHP (2017)[a]-VU	*Dendrobium*	Orchidaceae	Herb	Guangxi (Napo, Rongshui, Lingyun, Longzhou, Yongfu, Donglan, Jingxi, Longlin, etc.), south Guang-dong (Luofushan), Hainan (Baisha), Southwest Gui-zhou (Luodian, Xingyi, Guanling), South Yunnan (Simao, Mengla)	No	No	No	No	Yes	Subtropical monsoon	Forest
64	*Dendrobium williamsonii* Day et Reichb.F.	黑毛石斛	China RL (2013)-EN, China TSLHP (2017)[a]-EN	*Dendrobium*	Orchidaceae	Herb	Hainan (Wuzhishan and other places), northwest and north of Guangxi (Lingyun, Longlin, Rongshui, Donglan), South-east and west of Yunnan	No	No	No	No	Yes	Subtropical monsoon	Forest

65	*Dendrobium nobile* Lindl.	石斛	None	*Dendrobium*	Orchidaceae	Herb	Taiwan, South Hubei, Hong Kong, Hainan, West to North-east Guangxi, South Sichuan, southwest to North Guizhou, southeast to Northwest Yun-nan, Southeast Tibet	No	No	No	No	Yes	Tropical rainforest climate	Forest
66	*Diploknema yunnanensis*	滇藏榄(云南藏榄)	China RL (2013)-RE	*Diploknema*	Sapotaceae	Tree	Scattered in Southwest Yunnan	Yes	No	No	No	Yes	Subtropical monsoon and tropical monsoon climate	Forest
67	*Dipteronia dyeriana*	云南金钱槭	National key protected wild plants of China-II, China RL (2013)-EN, IUCN Red List-EN, China TSLHP (2017)[a] –EN, Acer RL-EN	*Dipteronia*	Aceraceae	Tree	Laojun Moun-tain, Wenshan County and Mingjiu District, Mengzi County, Southeast Yun-nan Province	Yes	Yes	No	No	Yes	Tropical monsoon	Forest
68	*Dipteronia sinensis*	金钱槭	China RL (2013)-LC, IUCN Red List-DD, Acer RL-NT	*Dipteronia*	Aceraceae	Tree	Southwest Henan, South Shaanxi, South-east Gansu, West Hubei, Sichuan, Guizhou, and other provinces	Yes	Yes	No	Yes	No	Temperate monsoon climate	Forest
69	*Disanthus cercidifolius* subsp. *longipes*	长柄双花木	National key protected wild plants of China-II, China RL (2013)-EN, IUCN Plants (1997)-E, China TSLHP (2017)[a] -EN	*Disanthus*	Hamamelidaceae	Tree	Junfeng Moun-tain in the east of Jiangxi Province and Changning and Dao County in Hunan Prov-ince, and Mang Mountain	Yes	No	No	Yes	Yes	Subtropical monsoon	Forest

(continued)

Table 1 (continued)

No.	Species	Chinese name	Endangered sp. listing (IUCN or China)	Genus	Family	Plant life form	Natural distribution (mainly based on Flora of China)	Chinese endemic	Endemic Genus	Endemic Family	Monogenous species	Tropical or subtropical species	Climate types of habitats	Ecosystem types of habitats
70	*Dunnia sinensis* Tutch.	绣球茜	China RL (2013)-LC, IUCN Plants (1997)-E	*Dunnia*	Rubiaceae	Shrub	Nankunshan, Guangdong Province	Yes	No	No	Yes	Yes	Subtropical monsoon or tropical rainforest climate	Forest
71	*Eleutharrhena macrocarpa*	藤枣	National key protected wild plants of China-I, China RL (2013)-CR, China TSLHP (2017)[a]-CR, IUCN Plants (1997)-E	*Eleutharrhena*	Menispermaceae	Vine	South and Southeast Yun-nan, China	Yes	No	No	Yes	Yes	Subtropical monsoon	Forest
72	*Emmenopterys henryi*	香果树	National key protected wild plants of China-II, China RL (2013)-NT, IUCN Plants (1997)-R	*Emmenopterys*	Rubiaceae	Tree	Shanxi, Gansu, Jiangsu, Anhui, Zhejiang, Jiangxi, Fujian, Henan, Hubei, Hunan, Guangxi, Sichuan, Gui-zhou, northeast-ern to central Yunnan	Yes	No	No	No	No	Temperate monsoon climate and temperate continental climate	Forest
73	*Eria corneri Rchb*	半柱毛兰	China RL (2013)-LC	*Eria*	Orchidaceae	Herb	South Fujian, Taiwan, Hainan, West Guang-dong, Hong Kong, South Guangxi, South-west Guizhou, and Southeast Yunnan	No	No	No	No	Yes	Subtropical monsoon	Forest

74	*Eriolaena kwangsiensis* Hand.-Mazz.	桂火绳 (广西芒木)	National key protected wild plants of China-II, China RL (2013)-EN, China TSLHP (2017)[a]-EN	*Eriolaena*	Sterculiaceae	Tree	Guangxi (Nanning, Donglan, Longzhou) and Yunnan (Lancang, Jingdong, Xishuangbanna)	Yes	No	No	No	Yes	Subtropical monsoon	Forest
75	*Erythropsis kwangsiensis*	广西火桐	National key protected wild plants of China-II	*Erythropsis*	Sterculiaceae	Tree	Jingxi, Napo, Tianyang, Fusui, Pingxiang, Dianqiangui, Guangxi	Yes	No	No	No	Yes	Subtropical monsoon	Forest
76	*Eucommia ulmoides*	杜仲	China RL (2013)-VU, China TSLHP (2017)[a]-VU, IUCN Red List-V	*Eucommia*	Eucommiaceae	Tree	Shaanxi, Gansu, Henan, Hubei, Sichuan, Yunnan, Guizhou, Hunan, and Zhejiang	Yes	Yes	Yes	Yes	Yes	Subtropical monsoon and temperate monsoon climate	Forest
77	*Eulophia bracteosa*	长苞美冠兰	China RL (2013)-VU, China TSLHP (2017)[a]-VU	*Eulophia*	Orchidaceae	Herb	North Guangdong (Liannan), Guangxi (Jinxiu, Xing'an) and Southeast Yunnan (Hekou, Pingbian)	No	No	No	No	Yes	Subtropical monsoon	Valley or bush
78	*Euptelea pleiosperma* Hook. f. et Thoms.	领春木	China RL (2013)-LC, IUCN Red List (2018)-LR/lc	*Euptelea*	Eupteleaceae	Tree	Hebei (Wu'an), Shaanxi (Yangcheng), Henan (Funiu Mountain), Shaanxi (Qinling), Gansu, Zhejiang (Tianmu Mountain), Hubei, Sichuan, Guizhou, Yunnan, Tibet	No	No	No	No	No	Subtropical monsoon and temperate monsoon climate	Forest

(continued)

Table 1 (continued)

No.	Species	Chinese name	Endangered sp. listing (IUCN or China)	Genus	Family	Plant life form	Natural distribution (mainly based on Flora of China)	Chinese endemic	Endemic Genus	Endemic Family	Monogenous species	Tropical or subtropical species	Climate types of habitats	Ecosystem types of habitats
79	*Eurycorymbus cavaleriei*	伞花木	China RL (2013)-LC, China TSLHP (2017)[a]-LC, IUCN (2014)-LR/nt	*Eurycorymbus*	Sapindaceae	Tree	Yunnan (Gongshan, Mengzi), Guizhou (guiding, Yinjiang, Zunyi, Xingyi, Xingren, Anlong, Dushan), Guangxi (Nandan, Xing'an, Guilin), Hunan (Yongsui, Huayuan, Chengdu), Jiangxi (Longnan, Anyuan), Guangdong (Lianxian, Yangshan, Wengyuan, Lechang, Pingyuan), Fujian (Changting, Longyan), Taiwan (Taipei, Hualian, Kaohsiung)	Yes	Yes	No	Yes	Yes	Subtropical monsoon and tropical monsoon climate	Forest
80	*Euryodendron excelsum*	猪血木	National key protected wild plants of China-II, China RL (2013)-CR, China TSLHP (2017)[a]-CR, IUCN (2014)-CR	*Euryodendron*	Theaceae	Tree	Bajia Town, Yangchun City, Guangdong Province	Yes	No	No	Yes	Yes	Subtropical monsoon	Sparse forest or forest edge

81	*Euryodendron excelsum* H. T. Chang	猪血木	China RL (2013)-CR, IUCN Red List-CR, China TSLHP (2017)[a] -CR	*Euryodendron*	Theaceae	Tree	Scattered in Bajia village, Yangchun, Guangdong, Siwang village, Pingnan, Guangxi and Linglu Township, Bama County	Yes	Yes	No	Yes	Yes	Subtropical monsoon	Forest
82	*Firmiana danxiaensis*	丹霞梧桐	China RL (2013)-CR, China TSLHP (2017)[a]-CR	*Firmiana*	Sterculiaceae	Tree	Danxiashan, Guangdong Province	Yes	No	No	No	Yes	Subtropical monsoon	Forest
83	*Fokienia hodginsii*	福建柏	China RL (2013)-VU, China TSLHP (2017)[a]-VU, IUCN (2014)-VU	*Fokienia*	Cupressaceae	Tree	South Zhejiang (Qingyuan, Longquan), Fujian (Daiyun Mountain, Yong'an, Yongtai, Zhangping, Changting, Xianyou in Dehua), North Guangdong (Lechang, Ruyuan), Jiangxi (Jingfeng Mountain), South Hunan (Mangshan in Yizhang), Guizhou (Dafang, Hezhang, Xishui, Dushan, Pingtang, Luodian),	No	No	No	Yes	Yes	Subtropical monsoon	Forest

(continued)

Table 1 (continued)

No.	Species	Chinese name	Endangered sp. listing (IUCN or China)	Genus	Family	Plant life form	Natural distribution (mainly based on Flora of China)	Chinese endemic	Endemic Genus	Endemic Family	Monogenous species	Tropical or subtropical species	Climate types of habitats	Ecosystem types of habitats
							Guangxi (Jinxiu, Longsheng), Sichuan (Jiangjin) and Southeast Yunnan (Maguan, Pingping) Bian, Pingshan, Dawei), and central Anning							
84	*Fosbergia shweliensis*	保山茜 (瑞丽茜树)	China RL (2013)-EN, China TSLHP (2017)[a]-EN	*Rhynchostylis*	Rubiaceae	Shrub	Western Yunnan Province	Yes	No	No	No	Yes	Subtropical monsoon	Forest
85	*Fraxinus hupehensis*	湖北梣 (湖北白蜡)	China RL (2013)-EN, China TSLHP (2017)[a]-EN	*Fraxinus*	Oleaceae	Tree	Hubei Province	Yes	No	No	No	No	Subtropical monsoon and temperate monsoon climate	Forest
86	*Fraxinus mandshurica*	水曲柳	IUCN Red List (2018)-LC, China TSLHP (2017)[a]-VU	*Fraxinus*	Oleaceae	Tree	Northeast, North China, Shaanxi, Gansu, Hubei, and other provinces	No	No	No	No	No	Temperate monsoon climate	Sparse forest on hillside or in gentle mountain valley
87	*Geodorum eulophioides*	贵州地宝兰	National key protected wild plants of China-II, China RL (2013)-EN, China TSLHP (2017)[a]-EN	*Geodorum*	Orchidaceae	Herb	Southern Guizhou (Luodian)	Yes	No	No	No	Yes	Subtropical monsoon	Forest

88	*Geodorum recurvum*	多花地宝兰	China RL (2013)-NT, IUCN (2014)-LC	*Geodorum*	Orchidaceae	Herb	South Guangdong, Hainan and south to Southeast Yunnan (Mengla, Shiping)	No	No	No	No	Yes	Subtropical monsoon	Forest
89	*Ginkgo biloba*	银杏	China RL (2013)-CR, China TSLHP (2017)[a]-CR, IUCN (2014)-EN	*Ginkgo*	Ginkgoaceae	Tree	Tianmu Mountain, Zhejiang Province	Yes	Yes	Yes	Yes	No	Subtropical monsoon and temperate monsoon climate	Forest
90	*Gleditsia vestita*	绒毛皂荚	IUCN (2014)-CR	*Gleditsia*	Leguminosae	Tree	Heng Mountain, Hunan	Yes	No	No	No	Yes	Subtropical monsoon	Sparse forest of mountain and roadside
91	*Glyptostrobus pensilis*	水松	National key protected wild plants of China-I, China RL (2013)-VU, China TSLHP (2017)[a]-CR, IUCN (2014)-CR	*Glyptostrobus*	Taxodiaceae	Tree	*Guangzhou Pearl River Delta, central Fujian and Lower Minjiang River*	Yes	Yes	No	Yes	Yes	Subtropical monsoon	Forest
92	*Habenaria dentata*	鹅毛玉凤花	China RL (2013)-LC	*Habenaria*	Orchidaceae	Herb	Anhui, Zhejiang, Jiangxi, Fujian, Taiwan, Hubei, Hunan, Guangdong, Guangxi, Sichuan, Guizhou, Yunnan, Tibet	No	No	No	No	Yes	Subtropical monsoon	Forest
93	*Heptacodium miconioides*	七子花	China RL (2013)-EN, China TSLHP (2017)[a]-EN, IUCN (2014)-VU	*Heptacodium*	Caprifoliaceae	Tree	Beixingshan County, Tiantai Mountain, Siming Mountain, Beishan Mountain of Yiwu, Tangjiawan of Changhua, Jingxian County, and Xuancheng of Anhui	Yes	Yes	No	Yes	Yes	Subtropical monsoon and temperate monsoon climate	Cliffs, hillside thickets, and forests

(continued)

Table 1 (continued)

No.	Species	Chinese name	Endangered sp. listing (IUCN or China)	Genus	Family	Plant life form	Natural distribution (mainly based on Flora of China)	Chinese endemic	Endemic Genus	Endemic Family	Monogenous species	Tropical or subtropical species	Climate types of habitats	Ecosystem types of habitats
94	*Juglans regia*	胡桃(核桃)	China RL (2013)-VU, China TSLHP (2017)[a]-VU, IUCN Red List (2019)-LC	*Juglans*	Juglandaceae	Tree	North, north-west, southwest, central, South, and East China	No	No	No	No	Yes	Subtropical monsoon and temper-ate mon-soon climate	Forest
95	*Keteleeria fortunei* (Murr) Carr.	油杉	China RL (2013)-LC, IUCN(2014)-NT, China TSLHP (2017)[a]-VU	*Keteleeria*	Pinaceae	Tree	Coastal moun-tains in southern Zhejiang, Fujian, Guangdong, and Guangxi	Yes	No	No	No	Yes	Subtropical monsoon	Coastal mountainous area
96	*Kolkwitzia amabilis*	蝟实(猬实)	China RL (2013)-VU, China TSLHP (2017)[a]-VU, IUCN Plants (1997)-R	*Kolkwitzia*	Caprifoliaceae	Shrub	Shanxi Province, Shaanxi Prov-ince, Gansu Province, Henan Province, Hubei Province, and Anhui Province	Yes	Yes	No	Yes	No	Temperate monsoon climate and temperate continental climate	Forest
97	*Lagerstroemia guilinensis*	桂林紫薇	China RL (2013)-EN	*Osmanthus*	Lythraceae	Shrub	Guangxi Guilin	Yes	No	No	No	Yes	Subtropical monsoon and temper-ate mon-soon climate	Shrub
98	*Ligustrum confusum Decne.*	散生女贞	IUCN Red List (2019)-LC	*Ligustrum*	Oleaceae	Shrub	Tibet of China	No	No	No	No	No	Subtropical monsoon	Forest
99	*Ligustrum expansum*	扩展女贞(扩叶女贞)	China RL (2013)-NT	*Ligustrum*	Oleaceae	Tree	Western Hubei Province	Yes	No	No	No	Yes	Subtropical monsoon	Forest

100	*Liparis viridiflora*	长茎羊耳蒜	China RL (2013)-LC	*Liparis*	Orchidaceae	Herb	Taiwan, Guangdong, Hainan, Guangxi, Southwest Sichuan, Yunnan, and Southeast Tibet (moto)	No	No	No	No	Yes	Subtropical monsoon	Forest
101	*Lirianthe fistulosa*	显脉木兰	None	*Lirianthe*	Magnoliaceae	Tree	Maguan County, Yunnan Province	No	No	No	No	Yes	Subtropical monsoon	Forest
102	*Liriodendron chinense*	鹅掌楸(马褂木)	National key protected wild plants of China-II, IUCN (2016)-NT, Magnolia RL-NT	*Liriodendron*	Magnoliaceae	Tree	Shaanxi (Zhenba), Anhui (Shexian, Xiuning, Shucheng, Yuexi, Qianshan, Huoshan), Zhejiang (Longquan, Suichang, Songyang), Jiangxi (Lushan), Fujian (Wuyishan), Hubei (Fangxian, Badong, Jianshi, Lichuan), Hunan (Sangzhi, Xinning), Guangxi (Rongshui, Lingui, Longsheng, Xing'an, resources, Guanyang, Huajiang), Sichuan (Wanyuan, Wan County, Xiushan,	No	No	No	No	No	Subtropical monsoon and temperate monsoon climate	Forest

(continued)

Table 1 (continued)

No.	Species	Chinese name	Endangered sp. listing (IUCN or China)	Genus	Family	Plant life form	Natural distribution (mainly based on Flora of China)	Chinese endemic	Endemic Genus	Endemic Family	Monogenous species	Tropical or subtropical species	Climate types of habitats	Ecosystem types of habitats
							Nanchuan, Xuyong, Gulin, Xilian, Guizhou (Suiyang, Xifeng, Liping), Yunnan (Yiliang, Daguan, Funing, Jinping, Malipo)							
103	*Litsea auriculata*	天目木姜子	China RL (2013)-VU, China TSLHP (2017)[a]-VU, IUCN (2014)-LR/nt	*Ormosia*	Lauraceae	Tree	Zhejiang (Tianmu Mountain and Tianhe Mountain), Southern Anhui (section county)	Yes	No	No	No	No	Subtropical monsoon and temperate monsoon climate	Forest
104	*Luisia teres*	叉唇钗子股	China RL (2013)-NT	*Luisia*	Orchidaceae	Herb	Taiwan (Wuli, Pingdong and Taidong), Western Guangxi (Xilin), central and Western Sichuan (Luding), Southwest Guizhou (Xingyi), Southeast Yunnan (Dananshan, Malipo)	No	No	No	No	Yes	Subtropical monsoon	Forest
105	*Magnolia amoena*	天目木兰	IUCN (2014)-VU, Magnolia RL-VU	*Magnolia*	Magnoliaceae	Tree	Zhejiang (Tianmu Mountain, Longquan, Suichang)	No	No	No	No	No	Subtropical monsoon	Forest

106	*Magnolia campbellii* (Hook. f. et Thoms.) D. L. Fu	滇藏木兰	IUCN (2014)-LC, Magnolia RL-LC	*Magnolia*	Magnoliaceae	Tree	Northwest (Weixi) and West (Ruili) of Yunnan and South Tibet	No	No	No	No	Yes	Subtropical monsoon	Forest
107	*Magnolia cylindrica*	黄山木兰	IUCN (2014)-VU, Magnolia RL-VU	*Magnolia*	Magnoliaceae	Tree	Anhui, Zhejiang, Jiangxi, Fujian, Southwest Hubei	Yes	No	No	No	Yes	Subtropical monsoon and temper-ate mon-soon climate	Mountain forest
108	*Magnolia insignis*	红花木莲	IUCN (2014)-LC, Magnolia RL-LC	*Magnolia*	Magnoliaceae	Tree	Null	No	No	No	No	Yes	Subtropical monsoon	Forest
109	*Magnolia odoratissima*	馨香木兰(馨香玉兰)	National key protected wild plants of China-II, IUCN (2016)-EN	*Magnolia*	Magnoliaceae	Tree	Guangnan, Yun-nan Province	Yes	No	No	No	Yes	Subtropical monsoon	Forest
110	*Magnolia officinalis*	厚朴	National key protected wild plants of China-II, IUCN (2016)-EN, Magnolia RL-NT, China RL (2013)-LC	*Magnolia*	Magnoliaceae	Tree	South Shaanxi, Southeast Gansu, Southeast Henan (Shangcheng, Xinxian), West Hubei, South-west Hunan, Sichuan (central, Eastern), north-east Guizhou	Yes	No	No	No	No	Subtropical monsoon	Forest
111	*Magnolia phanerophlebia* B. L. Chen	显脉木兰	None	*Magnolia*	Magnoliaceae	Tree	Null	No	No	No	No	Yes	Subtropical monsoon	Forest
112	*Magnolia sieboldii*	天女木兰(小花木兰)	None	*Magnolia*	Magnoliaceae	Tree	Liaoning, Anhui, Zhejiang, Jiangxi, Northern Fujian, Guangxi	No	No	No	No	No	Subtropical monsoon and temper-ate mon-soon climate	Forest
113				*Magnolia*	Magnoliaceae	Tree		Yes	No	No	No	Yes		Forest

(continued)

Table 1 (continued)

No.	Species	Chinese name	Endangered sp. listing (IUCN or China)	Genus	Family	Plant life form	Natural distribution (mainly based on Flora of China)	Chinese endemic	Endemic Genus	Endemic Family	Monogenous species	Tropical or subtropical species	Climate types of habitats	Ecosystem types of habitats
	Magnolia sinostellata P. L. Chiu et Z. H. Chen	景宁木兰	IUCN (2016)-EN				Western Guangxi (Napo), Southwest Guizhou (Xingyi) and Southeast Yunnan (Malipo, Wenshan, Maguan), China						Subtropical monsoon	
114	*Magnolia zenii*	宝华玉兰	IUCN (2014)-CR, Magnolia RL-CR	*Magnolia*	Magnoliaceae	Tree	Jiangsu (Jurong Baohua Mountain)	Yes	No	No	No	No	Subtropical monsoon and temperate monsoon climate	Forest
115	*Malania oleifera*	蒜头果	China RL (2013)-VU, China TSLHP (2017)[a]-VU, IUCN (2014)-VU	*Malania*	Olacaceae	Tree	Daxin, Longzhou, and Youjiang River Basin counties in Western Guangxi and Funing, Guangnan, and other places in eastern Yunnan	Yes	No	No	Yes	Yes	Tropical rainforest climate	Forest
116	*Malus ombrophila* Hand.-Mazz.	沧江海棠	China RL (2013)-NT, IUCN Red List-DD	*Malus*	Rosaceae	Tree	Northwest Yunnan, Southwest Sichuan	Yes	No	No	No	Yes	Subtropical monsoon	Forest

117	*Manglietia aromatica*	香木莲	National key protected wild plants of China-II, IUCN Plants (1997)-E, China RL (2013)-VU, China TSLHP (2017)[a] -VU	*Manglietia*	Magnoliaceae	Tree	The southeast of Yunnan Province and the west of Guangxi Zhuang Autonomous Region	No	No	No	No	Yes	Subtropical monsoon	Forest
118	*Manglietia longipedunculata*	长梗木莲	China RL (2013)-CR, China TSLHP (2017)[a] -CR, the Red List of Magnoliaceae	*Manglietia*	Magnoliaceae	Tree	The Nankunshan Nature Reserve in South China	Yes	No	No	No	Yes	Subtropical monsoon	Forest
119	*Manglietia patungensis*	巴东木莲	China RL (2013)-VU, China TSLHP (2017)[a]-VU	*Manglietia*	Magnoliaceae	Tree	West Hubei (Badong, Lichuan), South-east Sichuan (Hejiang, Nanchuan)	Yes	No	No	No	Yes	Subtropical monsoon	Forest
120	*Manglietia hebecarpa*	毛果木莲	National key protected wild plants of China-II	*Manglietia*	Magnoliaceae	Tree	Daweishan, Hekou and Jinping County, Pingbian County, Yunnan Province	Yes	No	No	No	Yes	Subtropical monsoon	Forest
121	*Mastixia euonymoides*	八蕊单室茱萸	None	*Cornaceae*	Mastixiaceae	Herb	Xishuangbanna, Pu'er, Yunnan	No	No	No	No	Yes	Tropical rainforest climate and tropical monsoon climate	Forest edge or sparse forest
122	*Metabriggsia ovalifolia*	单座苣苔	National key protected wild plants of China-I, China RL (2013)-NT	*Metabriggsia*	Gesneriaceae	Herb	Napo, Guangxi	Yes	No	No	No	Yes	Subtropical monsoon	Forest

(continued)

Table 1 (continued)

No.	Species	Chinese name	Endangered sp. listing (IUCN or China)	Genus	Family	Plant life form	Natural distribution (mainly based on Flora of China)	Chinese endemic	Endemic Genus	Endemic Family	Monogenous species	Tropical or subtropical species	Climate types of habitats	Ecosystem types of habitats
123	*Metasequoia glyptostroboides*	水杉	China RL (2013)-EN, IUCN (2014)-EN, China TSLHP (2017)[a]-CR	*Metasequoia*	Taxodiaceae	Tree	East Sichuan (Shizhu County), Southwest Hubei (Lichuan), and Northwest Hunan (Longshan and Sangzhi) mountainous areas	Yes	Yes	No	1子遗种	No	Subtropical monsoon and temperate monsoon climate	Forest
124	*Metasequoia glyptostroboides*	水杉	China RL (2013)-EN, China TSLHP (2017)[a]-CR, IUCN (2014)-EN	*Metasequoia*	Taxodiaceae	Tree	Shizhu County, Sichuan Province, and Mocaoxi County, Lichuan County, Hubei Province	Yes	No	No	Yes	Yes	Subtropical monsoon	Forest
125	*Michelia coriacea* Chang et B. L. Chen	西畴含笑	None	*Michelia*	Magnoliaceae	Tree	Xichou County, Yunnan Province	Yes	No	No	No	Yes	Null	Null
126	*Michelia odora* (Chun) Nooteboom & B. L. Chen	观光木	China RL (2013)-VU, China TSLHP (2017)[a]-VU	*Michelia*	Magnoliaceae	Tree	South Jiangxi, Fujian, Guangdong, Hainan, Guangxi, Southeast Yunnan	Yes	Yes	No	Yes	Yes	Subtropical monsoon and temperate monsoon climate	Forest
127	*Michelia wilsonii*	峨眉含笑	IUCN Plants (1997)-E	*Michelia*	Magnoliaceae	Tree	Central and Western Sichuan	Yes	No	No	No	Yes	Subtropical monsoon and temperate monsoon climate	Forest

128	*Myricaria laxiflora*	疏花水柏枝	China RL (2013)-EN, China TSLHP (2017)[a]-EN	*Myricaria*	Tamaricaceae	Shrub	Zigui, Badong, Hubei, and Wushanxiakou, Sichuan	Yes	No	No	No	Yes	Subtropical monsoon and temperate monsoon climate	Wetland
129	*Myrsine semiserrata* Wall.	针齿铁仔	China RL (2013)-LC, IUCN Red List (2019)-LC	*Myrsine*	Myrsinaceae	Shrub	Hubei Province, Hunan Province, Guangxi Province, Guangdong Province, Sichuan Province, Guizhou Province, Yunnan Province, Tibet Province,	No	No	No	No	Yes	Subtropical monsoon	Forest
130	*Neolitsea sericea*	舟山新木姜子	China RL (2013)-EN, China TSLHP (2017)[a]-EN	*Neolitsea*	Lauraceae	Tree	Zhejiang (Zhoushan) and Shanghai (Chongming Island)	No	No	No	No	Yes	Subtropical monsoon	Forest
131	*Nothodoritis zhejiangensis*	象鼻兰	China RL (2013)-EN, China TSLHP (2017)[a]-EN	*Nothodoritis*	Olacaceae	Herb	Tianmu Mountain in Lin'an and Ningbo,Zhejiang Province	Yes	No	No	Yes	Yes	Subtropical monsoon and tropical monsoon climate	Forest or forest edge
132	*Nypa fruticans*	水椰	China RL (2013)-VU, IUCN (2014)-LC, China TSLHP (2017)[a] (2017)-VU	*Nypa*	Palmae	Palm	The mudflat zone of coastal harbors in Yaxian, Lingshui, Wanning, Wenchang, and other counties in the southeast of Hainan Province	No	No	No	No	Yes	Tropical monsoon	Forest

(continued)

Table 1 (continued)

No.	Species	Chinese name	Endangered sp. listing (IUCN or China)	Genus	Family	Plant life form	Natural distribution (mainly based on Flora of China)	Chinese endemic	Endemic Genus	Endemic Family	Monogenous species	Tropical or subtropical species	Climate types of habitats	Ecosystem types of habitats
133	*Nyssa yunnanensis*	云南蓝果树	China RL (2013)-CR, China TSLHP (2017)[a]-CR, IUCN(2014)-CR	*Nyssa*	Nyssaceae	Tree	Xishuangbanna, Yunnan Province	Yes	No	No	No	Yes	Subtropical monsoon or tropical rainforest climate	Forest
134	*Oberonia myosurus*	棒叶鸢尾兰	None	*Oberonia*	Orchidaceae	Herb	South to south-west of Guizhou (Huishui, Xingyi), west of Guangxi (Lingyun), South to southeast of Yunnan (Simao, Jianshui, Menghai, Mengla, Yanshan)	No	No	No	No	Yes	subtropical monsoon	Forest
135	*Ormosia hosiei*	红豆树	National key protected wild plants of China-II、China RL (2013)-EN, China TSLHP (2017)[a]-EN, IUCN(2014)-LR/nt	*Ormosia*	Leguminosae	Tree	Shaanxi (South), Gansu (south-east), Jiangsu, Anhui, Zhejiang, Jiangxi, Fujian, Hubei, Sichuan, Guizhou.	Yes	No	No	No	Yes	Temperate monsoon climate and subtropical monsoon	Riverside/ forest
136	*Oryza rufipogon*	野生稻	China RL (2013)-CR, China TSLHP (2017)[a]-RE, IUCN(2014)-CR	*Oryza*	Gramineae	Herb	Guangdong, Hai-nan, Guangxi, Yunnan, Taiwan	No	No	No	No	Yes	Subtropical monsoon and tropical monsoon climate	Wetland

137	*Osmanthus yunnanensis* (Franch.) P. S. Green	野桂花 (云南桂花)	China RL (2013)-LC	*Osmanthus*	Oleaceae	Shrub	Yunnan, Sichuan, Tibet, and other places	No	No	No	No	Yes	Plateau mountain climate	Forest
138	*Ostrya rehderiana*	天目铁木	China RL (2013)-CR, China TSLHP (2017)[a]-CR, IUCN (2014)-CR	*Ostrya*	Betulaceae	Tree	Tianmu Mountain, Zhejiang Province	Yes	No	No	No	Yes	Subtropical monsoon	Forest
139	*Ostrya rehderiana*	天目铁木	National key protected wild plants of China-I, China RL (2013)-CR, China TSLHP (2017)[a]-CR, IUCN(2014)-CR	*Ostrya*	Betulaceae	Tree	West Tianmu Mountain, Zhejiang Province	Yes	No	No	No	Yes	Subtropical monsoon climate	Forest edge or forest side
140	*Pachylarnax sinica*	华盖木	National key protected wild plants of China-I, China RL (2013)-CR, China TSLHP (2017)[a]-CR	*Pachylarnax*	Magnoliaceae	Tree	Yunnan Province (Xichou fadou)	Yes	No	No	Yes	Yes	Subtropical monsoon	Forest
141	*Paeonia suffruticosa* var. *papaveracea*	紫斑牡丹	IUCN Plants (1997)-V	*Paeonia*	Ranunculaceae	Shrub	Northern Sichuan, southern Gansu, southern Shaanxi (Taibai Mountain Area)	Yes	No	No	No	No	Temperate monsoon climate	Sparse forest on hillside or in gentle mountain valley
142	*Paphiopedilum armeniacum*	杏黄兜兰	China RL (2013)-CR, IUCN (2016)-EN, China TSLHP (2017)[a]-CR	*Paphiopedilum*	Orchidaceae	Herb	Western Yunnan (Bijiang, Lushui)	No	No	No	No	Yes	Plateau mountain climate	Forest

(continued)

Table 1 (continued)

No.	Species	Chinese name	Endangered sp. listing (IUCN or China)	Genus	Family	Plant life form	Natural distribution (mainly based on Flora of China)	Chinese endemic	Endemic Genus	Endemic Family	Monogenous species	Tropical or subtropical species	Climate types of habitats	Ecosystem types of habitats
143	*Paphiopedilum hirsutissimum* (Lindl. et Hook.) Stein	带叶兜兰	China RL (2013)-VU, China TSLHP (2017)[a]-VU, IUCN(2016)-VU	*Paphiopedilum*	Orchidaceae	Herb	From west to north of Guangxi (Longzhou, Tian'e), south-west of Guizhou (Xingyi, etc.), and southeast of Yunnan (Funing, Wenshan, Malipo)	No	No	No	No	Yes	Subtropical monsoon	Forest
144	*Paphiopedilum malipoense* S. C. Chen & Z. H. Tsi	麻栗坡兜兰	IUCN(2016)-EN	*Paphiopedilum*	Orchidaceae	Herb	Malipo County, Wenshan Prefec-ture, Yunnan Province	No	No	No	No	Yes	Subtropical monsoon	Forest
145	*Paphiopedilum micranthum* T. Tang et F. T. Wang	硬叶兜兰	China RL (2013)-VU, China TSLHP (2017)[a]-VU	*Paphiopedilum*	Orchidaceae	Herb	Southwest Guangxi, South and southwest Guizhou (Libo, Xingyi), and Southeast Yun-nan (Malipo, Xichou, Wenshan)	No	No	No	No	Yes	Subtropical monsoon	Forest
146	*Paphiopedilum purpuratum*	紫纹兜兰	China RL (2013)-EN, China TSLHP (2017)[a]-EN, IUCN(2016)-CR	*Paphiopedilum*	Olacaceae	Herb	Southern Guang-dong (Yangchun), Hong Kong, southern Guangxi (Shangsi 100,000 Dashan), and Southeast Yun-nan (Wenshan)	No	No	No	No	Yes	Tropical rainforest climate	Forest

147	*Paphiopedilum spicerianum*	白旗兜兰	China RL (2013)-CR, China TSLHP (2017)[a]-CR, IUCN(2016)-EN	*Papilionanthe*	Orchidaceae	Herb	Yunnan (Gaoligong Mountain, Simao)	No	No	No	No	Yes	Subtropical monsoon and tropical monsoon climate	Wetland/ grassland
148	*Paphiopedilum wardii*	彩云兜兰	China RL (2013)-DD, IUCN(2016)-EN	*Paphiopedilum*	Orchidaceae	Herb	Lushui County and Northwest of Baoshan City, Western Yunnan, China and Northern Myanmar	No	No	No	No	Yes	Tropical monsoon	Forest
149	*Paphiopedilum dianthum*	长瓣兜兰	China RL (2013)-VU, China TSLHP (2017)[a]-VU, IUCN(2016)-EN	*Paphiopedilum*	Orchidaceae	Herb	Southwest Guangxi (Jingxi), Southwest Guizhou (Xingyi), and Southeast Yunnan (Malipo)	No	No	No	No	Yes	Subtropical monsoon	Forest
150	*Papilionanthe teres*	凤蝶兰	China RL (2013)-VU, China TSLHP (2017)[a]-VU	*Papilionanthe*	Orchidaceae	Herb	Southern Yunnan Province (Mengla)	No	No	No	No	Yes	Tropical rainforest climate and tropical monsoon climate	Forest
151	*Parakmeria omeiensis*	峨眉拟单性木兰	National key protected wild plants of China-I	*Parakmeria*	Magnoliaceae	Tree	Mount Emei, Sichuan Province	Yes	No	No	No	Yes	Subtropical monsoon	Forest
152	*Paris luquanensis*	禄劝花叶重楼	None	*Paris*	Liliaceae	Herb	yi, WeishanYunnan Province, and Huili, Puge, Yuexi, Sichuan Province	Yes	No	No	No	Yes	Subtropical monsoon and tropical monsoon climate	Forest
153	*Parrotia subaequalis*	银缕梅	National key protected wild plants of China-I	*Parrotia*	Hamamelidaceae	Shrub	In Jiangsu, Zhejiang, Anhui, Chongqing, Henan, and other deep mountains	Yes	No	No	No	Yes	Subtropical monsoon	Forest

(continued)

Table 1 (continued)

No.	Species	Chinese name	Endangered sp. listing (IUCN or China)	Genus	Family	Plant life form	Natural distribution (mainly based on Flora of China)	Chinese endemic	Endemic Genus	Endemic Family	Monogenous species	Tropical or subtropical species	Climate types of habitats	Ecosystem types of habitats
154	*Phaius flavus* (Bl.) Lindl.	黄花鹤顶兰	China RL (2013)-LC	*Phaius*	Orchidaceae	Herb	Fujian (Longxi, Jianyang, Chongan), Taiwan, Northern Hunan (Shimen), Guangdong (Luofushan, Raoping, Wengyuan, HUAIJI, Xinyi, Ruyuan, Lianxian, Lechang), Guangxi (Pingnan, Rongshui, Xing'an), Hong Kong, Hainan, Guizhou (Fanjingshan), Sichuan (Leibo, Tianquan, Junlian, Hejiang, Nanchuan, Emeishan, etc.), Yunnan (Malipo, Funing, Jinghong, Weixi and Nujiang River Valley) and Southeast, Tibet (Motuo)	No	No	No	No	Yes	Subtropical monsoon	Forest

155	*Phellodendron amurense*	黄檗	China RL (2013)-VU, China TSLHP (2017)[a]-VU	*Phellodendron*	Rutaceae	Tree	Northeast and North China provinces, Henan, Northern Anhui	No	No	No	No	No	Temperate monsoon climate	Forest
156	*Phoebe bournei*	闽楠	China RL (2013)-VU, China TSLHP (2017)[a]-VU, IUCN(2014)-LR/nt	*Phoebe*	Lauraceae	Tree	Jiangxi Province, Fujian Province, South Zhejiang Province, Guangdong Province, North and Northeast Guangxi Province, Hunan Province, Hubei Province, Southeast and northeast Guizhou Province	No	No	No	No	Yes	Subtropical monsoon	Forest
157	*Phoebe chekiangensis*	浙江楠	National key protected wild plants of China-II, China RL (2013)-VU, IUCN (2014)-VU, China TSLHP (2017)[a]-VU	*Phoebe*	Lauraceae	Tree	Northwest and northeast Zhejiang, North Fujian and East Jiangxi	Yes	No	No	No	No	Subtropical monsoon and temperate monsoon climate	Forest
158	*Pholidota yunnanensis*	云南石仙桃	China RL (2013)-NT	*Pholidota*	Orchidaceae	Herb	Guangxi, Western Hubei, Western Hunan, northeast to South Sichuan, Guizhou, and Southeast Yunnan	No	No	No	No	Yes	Subtropical monsoon	Forest

(continued)

Table 1 (continued)

No.	Species	Chinese name	Endangered sp. listing (IUCN or China)	Genus	Family	Plant life form	Natural distribution (mainly based on Flora of China)	Chinese endemic	Endemic Genus	Endemic Family	Monogenous species	Tropical or subtropical species	Climate types of habitats	Ecosystem types of habitats
159	*Pinus dabeshanensis*	大别山五针松	IUCN Plant (1997)-E	*Pinus*	Pinaceae	Tree	Dabie Mountains in Southwest Anhui (Yuexi) and East Hubei (Yingshan and Luotian)	Yes	No	No	No	No	Subtropical monsoon	Forest
160	*Pinus fenzeliana* Hand.-Mazz. var. *dabeshanensis* (C. Y. Cheng et Y. W. Law) L. K. Fu et Nan Li	大别山五针松	National key protected wild plants of China-II, *Pinus fenzeliana*	*Pinus*	Pinaceae	Tree	Dabie Mountains in Southwest Anhui (Yuexi) and East Hubei (Yingshan and Luotian)	Yes	No	No	No	Yes	Subtropical monsoon	Forest
161	*Pinus squamata*	巧家五针松	National key protected wild plants of China-I, China RL (2013)-CR, China TSLHP (2017)[a]-CR, IUCN(2014)-CR	*Pinus*	Pinaceae	Tree	Qiaojia County, Yunnan Province	Yes	No	No	No	Yes	Subtropical monsoon	Forest
162	*Podocarpus fleuryi*	长叶竹柏	IUCN(2014)-NT	*Podocarpus*	Pinaceae	Tree	Mengzi, Dawei Mountain Area in Pingbian, Hepu, Guangxi, Gaoyao, Zengcheng, Longmen, and other places in Southeast Yunnan	No	No	No	No	Yes	Subtropical monsoon	Forest

163	*Podocarpus nagi*	竹柏	None	*Podocarpus*	Podocarpaceae	Tree	Zhejiang, Fujian, Jiangxi, Hunan, Guangdong, Guangxi, Sichuan	No	No	No	No	Yes	Tropical rainforest climate	Forest
164	*Poncirus polyandra*	富民枳	None	*Poncirus*	Rutaceae	Tree	Long Ma Cun Long Tan Qing Lao Qing Shan, Fumin County, Kunming City, Yunnan Province	Yes	No	No	No	No	Plateau mountain climate	Forest
165	*Primulina tabacum*	报春苣苔	National key protected wild plants of China-I, China RL (2013)-EN, China TSLHP (2017)[a] -EN	*Primulina*	Gesneriaceae	Herb	The entrances of karst cave drainages along the border between northern Guangdong and southern Hunan, China	Yes	No	No	No	Yes	Central subtropical monsoon	Other (karst cave)
166	*Pseudolarix amabilis*	金钱松	China RL (2013)-VU, China TSLHP (2017)[a]-VU	*Pseudolarix*	Pinaceae	Tree	South Jiangsu (Yixing), South Zhejiang, South Anhui, North Fujian, Jiangxi, Hunan, Hubei Lichuan to Wanxian County, Sichuan	Yes	No	No	No	No	Temperate monsoon climate	Forest
167	*Pseudotsuga sinensis*	黄杉	China TSLHP (2017)[a]-VU, IUCN(2014)-VU	*Pseudotsuga*	Pinaceae	Tree	Yunnan, Sichuan, Guizhou, Hubei, Hunan	Yes	No	No	No	Yes	Subtropical monsoon	Forest
168	*Psilopeganum sinense*	裸芸香	China RL (2013)-EN, China TSLHP (2017)[a]-EN	*Psilopeganum*	Rutaceae	Herb	Northwest Hubei, Northeast Sichuan, Guizhou (Chishui)	Yes	Yes	No	Yes	Yes	Subtropical monsoon	Forest

(continued)

Table 1 (continued)

No.	Species	Chinese name	Endangered sp. listing (IUCN or China)	Genus	Family	Plant life form	Natural distribution (mainly based on Flora of China)	Chinese endemic	Endemic Genus	Endemic Family	Monogenous species	Tropical or subtropical species	Climate types of habitats	Ecosystem types of habitats
169	*Pterocarya stenoptera* var. *zhijiangensis*	枝江枫杨	None	*Pterocarya*	Juglandaceae	Tree	Zhijiang City, Hubei Province	Yes	No	No	No	Yes	Subtropical monsoon	Forest
170	*Pterostyrax psilophyllus*	白辛树	None	*Pterostyrax*	Styracaceae	Tree	Hunan Province (Yizhang, Xinning, Sangzhi, Yongshun), Hubei Province (Badong, Yichang), Sichuan Province (Jinshan), Guizhou Province (Kaili, Pingfa, Duyun, Anlong, Wangmo), Guangxi Province (Quanzhou, Xing'an, Longlin), and Yunnan Province (Zhenxiong)	No	No	No	No	Yes	Subtropical monsoon	Forest
171	*Ranalisma rostratum*	长喙毛茛泽泻	China RL (2013)-CR, China TSLHP (2017)[a]-CR	*Ranalisma*	Alismataceae	Herb	Siming Mountain, Lishui, Zhejiang Province	No	No	No	No	Yes	Subtropical monsoon and tropical monsoon climate	Wetland
172	*Rhododendron argyrophyllum* Franch. subsp. *argyrophyllum*	银叶杜鹃	China RL (2013)-LC, Rhododendron RL-LC	*Rhododendron*	Ericaceae	Shrub	South Sichuan, East Guizhou	No	No	No	No	Yes	Subtropical monsoon	Forest

173	*Rhynchostylis retusa*	钻喙兰	China RL (2013)-EN, China TSLHP (2017)[a]-EN	*Rhynchostylis*	Orchidaceae	Herb	Southwest of the prefecture (Xingyi), south-east to southwest of Yunnan (Jinping, Malipo, Pingbian, Mengla, Menghai, Jinghong, Cangyuan, Simao, Zhenkang)	No	No	No	No	Yes	Subtropical monsoon or tropical rainforest climate	Forest
174	*Robiquetia succisa*	寄树兰	China RL (2013)-LC	*Robiquetia*	Orchidaceae	Herb	Fujian (Nanjing, Pinghe, Hua'an, Yongtai, Lianjiang, Shunchang, etc.), western Guang-dong (Huaiji), Hong Kong, Hainan (Lingshui, Changjiang, Danxian), East-ern Guangxi (Cangwu), southern Yunnan (Mengla, Menghai, Jinghong)	No	No	No	No	Yes	Subtropical monsoon	Forest
175	*Sarcococca wallichii* Stapf.	云南野扇花(厚叶清香桂)	China RL (2013)-LC	*Sarcococca*	Buxaceae	Shrub	Dali, Jingdong, Zhenkang, Tengchong, and other counties in Yunnan Province	No	No	No	No	Yes	Tropical monsoon	Forest

(continued)

Table 1 (continued)

No.	Species	Chinese name	Endangered sp. listing (IUCN or China)	Genus	Family	Plant life form	Natural distribution (mainly based on Flora of China)	Chinese endemic	Endemic Genus	Endemic Family	Monogenous species	Tropical or subtropical species	Climate types of habitats	Ecosystem types of habitats
176	*Schima argentea*	银木荷	China RL (2013)-LC, IUCN Red List (2019)-LC	*Schima*	Theaceae	Tree	Sichuan, Yunnan, Guizhou, Hunan	No	No	No	No	Yes	Subtropical monsoon	Forest
177	*Shorea assamica*	阿萨姆娑罗双(云南娑罗双)	China RL (2013)-EN, China TSLHP (2017)[a]-EN, IUCN Red List (2019)-LC	*Shorea*	Dipterocarpaceae	Tree	Mengnai River Basin, Tongbiguan Nature Reserve, Yingjiang(盈江铜壁关自然保护区勐乃河流域)	No	No	No	No	Yes	Tropical rainforest climate	Forest
178	*Sinojackia dolichocarpa*	长果秤锤树	China RL (2013)-EN, IUCN (2014)-VU, China TSLHP (2017)[a]-EN	*Sinojackia*	Styracaceae	Tree	Shimen, Hunan Province	Yes	Yes	No	No	No	Subtropical monsoon and temperate monsoon climate	Forest
179	*Sinojackia dolichocarpa*	长果秤锤树(长果安息香)	National key protected wild plants of China-II	*Sinojackia*	Sterculiaceae	Tree	Tianmen Mountain, Zhangjiajie, Hunan, Badagong Mountain, Humen Huping Mountain and Maoping Town, Zigui County, Hubei Province	Yes	No	No	No	Yes	Subtropical monsoon	Wetland
180	*Sinojackia microcarpa*	细果秤锤树	National key protected wild plants of China-II, China RL (2013)-CR, China TSLHP (2017)[a]-CR	*Sinojackia*	Styracaceae	Tree	Lin'an and Jiande, Zhejiang Province	Yes	Yes	No	No	Yes	Subtropical monsoon and tropical monsoon climate	Forest

181	*Sinowilsonia henryi*	山白树	IUCN(2014)-LR/nt	*Sinowilsonia*	Hamamelidaceae	Tree	Hubei, Sichuan, Henan, Shaanxi, and Gansu	Yes	Yes	No	Yes	No	Temperate monsoon climate and temperate continental climate	Forest
182	*Styrax zhejiangensis*	浙江安息香	China RL (2013)-CR, China TSLHP (2017)[a]-CR	*Styrax*	Styracaceae	Tree	Jiande City, Zhejiang Province	Yes	No	No	No	Yes	Subtropical monsoon and tropical monsoon climate	Waterside
183	*Tacca chantrieri Andre*	箭根薯	China RL (2013)-NT	*Tacca*	Taccaceae	Herb	Southern Hunan, Guangdong, Guangxi, Yunnan	No	No	No	No	Yes	Subtropical monsoon and tropical monsoon climate	Forest
184	*Taiwania flousiana*	台湾杉	IUCN Plants (1997)-R	*Taiwania*	Taxodiaceae	Tree	Taiwan's central mountains	Yes	No	No	No	Yes	Subtropical monsoon	Forest
185	*Tapiscia sinensis*	银鹊树(瘿椒树)	IUCN (2014)-VU	*Tapiscia*	Staphyleaceae	Tree	Zhejiang, Anhui, Hubei, Hunan, Guangdong, Guangxi, Sichuan, Yunnan, Guizhou. Living mountain forest	Yes	Yes	No	No	Yes	Subtropical monsoon	Forest
186	*Taxus yunnanensis* Cheng et L. K. Fu	云南红豆杉	IUCN Plants (1997)-V	*Taxus*	Taxaceae	Tree	Northwest and West Yunnan (Zhenkang, Jingdong), Southwest Sichuan, and Southeast Tibet	No	No	No	No	No	Plateau mountain climate	Forest
187	*Tetracentron sinense* Oliv.	水青树	National key protected wild plants of China-II, China RL (2013)-LC	*Tetracentron*	Tetracentraceae	Tree	Northwest, northeast, Longling, Fengqing, Jingdong, Wenshan, Jinping	No	No	No	No	Yes	Subtropical monsoon	Forest

(continued)

Table 1 (continued)

No.	Species	Chinese name	Endangered sp. listing (IUCN or China)	Genus	Family	Plant life form	Natural distribution (mainly based on Flora of China)	Chinese endemic	Endemic Genus	Endemic Family	Monogenous species	Tropical or subtropical species	Climate types of habitats	Ecosystem types of habitats
188	*Tetradium daniellii* (Bennett) T. G. *Hartley*	臭檀吴萸	National key protected wild plants of China-II, China RL (2013)-LC, IUCN Red List (2019)-LC	*Tetradium*	Rutaceae	Tree	South of Liaoning Province to the Yangtze River	No	No	No	No	Yes	Subtropical monsoon	Forest
189	*Tetrathyrium subcordatum*	四药门花	National key protected wild plants of China-II, IUCN Plants (1997)-R	*Tetrathyrium*	Hamamelidaceae	Shrub	Wuguishan Nature Reserve in Guangdong Province, Maolan National Nature Reserve in Guizhou Province, Lantau Island and Baoyun road in Hong Kong	Yes	Yes	No	Yes	Yes	Subtropical monsoon	Forest
190	*Thuja sutchuenensis Franch.*	崖柏	China RL (2013)-EN, China TSLHP (2017)[a]-EN, IUCN(2014)-EN	*Thuja*	Cupressaceae	Tree	Daba Mountain in Sichuan and Taihang Mountain in North China	No	No	No	No	Yes	Tropical rainforest climate	Forest
191	*Tigridiopalma magnifica* C. Chen	虎颜花	China RL (2013)-EN, China TSLHP (2017)[a] -EN	*Tigridiopalma*	Melastomataceae	Herb	Southern Guangdong Province	Yes	Yes	No	Yes	Yes	Subtropical monsoon	Forest/ wetland
192	*Toona ciliata* var. *pubescens*	毛红椿	National key protected wild plants of China-II, China RL (2013)-EN, China TSLHP (2017)[a]-EN	*Toona*	Meliaceae	Tree	Fujian, Hunan, Guangdong, Guangxi, Sichuan, and Yunnan	No	No	No	No	Yes	Subtropical monsoon	Forest

193	*Torreya grandis* var. *jiulongshanensis* Z. Y. Li et al.	九龙山榧	China RL (2013)-CR, IUCN(2014)-DD	*Torreya*	Taxaceae	Tree	Jiulongshan National Nature Reserve	Yes	No	No	No	Yes	Subtropical monsoon and tropical monsoon climate	Forest
194	*Trigonobalanus doichangensis*	三棱栎	National key protected wild plants of China-II, IUCN Red List (2019)-E	*Trigonobalanus*	Orchidaceae	Tree	Lancang, Menglian, Ximeng, and other counties in Yunnan, China	No	No	No	No	Yes	Subtropical monsoon	Forest
195	*Tsuga chinensis* var. *chekiangensis*	南方铁杉	IUCN Plants (1997)-V	*Tsuga*	Pinaceae	Tree	Bailong River Basin in Gansu, South Shaanxi, West Henan, West Hubei, Northeast Sichuan and upper reaches of Minjiang River Basin, big and small Jinchuan River Basin, Dadu River Basin, Qingyi River Basin, lower reaches of Jinsha River Basin and Northwest Guizhou	Yes	No	No	No	No	Subtropical monsoon	Forest
196	*Tsuga forrestii*	丽江铁杉	IUCN (2014)-VU	*Tsuga*	Pinaceae	Tree	Northwest Yunnan (Lijiang, Zhongdian), Southwest Sichuan (Muli, Jiulong, Dechang)	Yes	No	No	No	Yes	Subtropical monsoon	Forest
197	*Ulmus gaussenii*	醉翁榆		*Ulmus*	Ulmaceae	Tree		Yes	No	No	No	No		Forest

(continued)

Table 1 (continued)

No.	Species	Chinese name	Endangered sp. listing (IUCN or China)	Genus	Family	Plant life form	Natural distribution (mainly based on Flora of China)	Chinese endemic	Endemic Genus	Endemic Family	Monogenous species	Tropical or subtropical species	Climate types of habitats	Ecosystem types of habitats
			China RL (2013)-CR, China TSLHP (2017)[a]-CR, IUCN (2014)-CR				Langya Mountain, Chuxian County, Anhui Province						Subtropical monsoon and temperate monsoon climate	
198	*Ulmus chenmoui*	琅牙榆	None	*Ulmus*	Ulmaceae	Tree	Null	Yes	No	No	No	Yes	Subtropical monsoon	Forest
199	*Vandopsis gigantea*	拟万代兰	China RL (2013)-LC	*Vandopsis*	Orchidaceae	Herb	Southwest Guangxi (Longzhou), South Yunnan (Mengla, Menghai)	No	No	No	No	Yes	Subtropical monsoon	Forest
200	*Zanthoxylum oxyphyllum* Edgew.	尖叶花椒	China RL (2013)-LC	*Zanthoxylum*	Rutaceae	Shrub	The west of Yunnan Province and Nyalam and Zana in Tibet	No	No	No	No	No	Plateau mountain climate	Forest
201	*Zenia insignis*	任豆	National key protected wild plants of China-II, China RL (2013)-VU, IUCN (2018)-LC, China TSLHP (2017)[a]-VU	*Zenia*	Leguminosae	Tree	Guangdong Province, Guangxi Province	No	Yes	No	Yes	Yes	Subtropical monsoon	Forest
202	*Taxus cuspidata* Siebold & Zucc	东北红豆杉	National key protected wild plants of China-I, IUCN-CR	*Taxus*	Taxaceae	Tree	Jilin Province	No	No	No	No	No	Temperate continental mountain climate	Forest

203	*Acer catalpifolium* Rehd	梓叶槭	National key protected wild plants of China-II	*Acer*	Aceraceae	Tree/shrub	Sichuan Province	Yes	No	No	No	Yes	Subtropical monsoon	Forest
204	*Betula halophila* Ching	盐桦	National key protected wild plants of China-II, IUCN-EX	*Betula*	Betulaceae	Shrub	Xingjiang Autonomous Region	Yes	No	No	No	No	Temperate continental mountain climate	Salt marsh
205	*Hopea hainanensis* Merr. et Chun	坡垒	China RL (2013)-EN, IUCN-EN	*Hopea*	Dipterocarpaceae	Tree	Hainan, Yunnan Province	No	No	No	No	Yes	Tropical monsoon	Forest
206	*Dayaoshania cotinifolia W.T. Wang*	瑶山苣苔	National key protected wild plants of China-I, IUCN-CR	*Dayaoshania*	Gesneriaceae	Grass	Guangxi Autonomous Region	Yes	No	No	No	Yes	Subtropical monsoon	Forest

[a]China TSLHP (2017), Threatened Species List of China's Higher Plants (Qin et al. 2017)

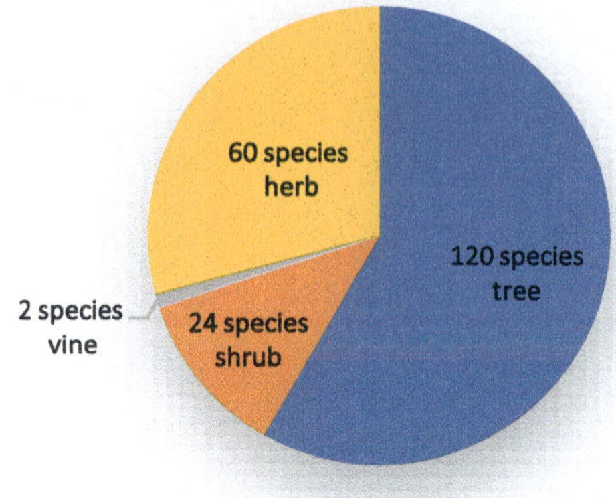

Fig. 1 Some rare and endangered plants reintroduced

Fig. 2 Some Magnoliaceae plants reintroduced in China

Recommendations for Chinese wild plant conservation and reintroduction are as follows: (1) Revise laws and regulations on wild plant protection, update the red list, and strictly implement laws and regulations. (2) Systematically summarize the past

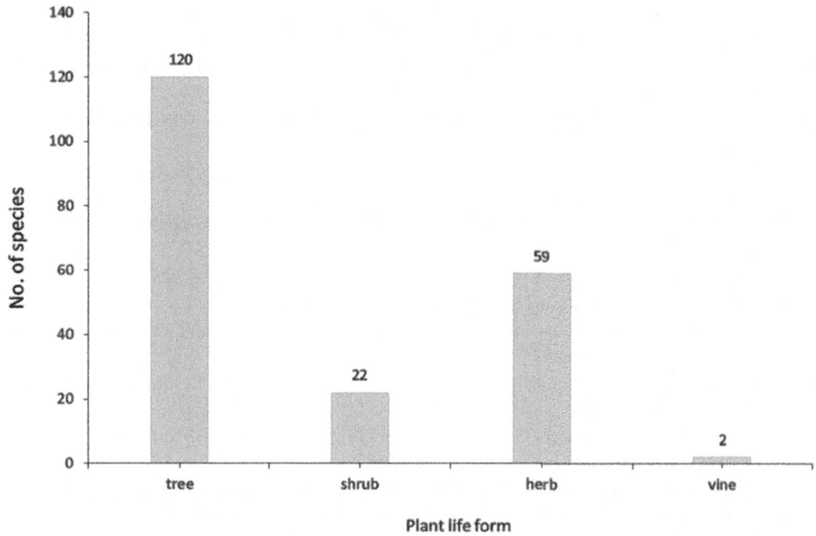

Fig. 3 Number of conservation translocation projects in China relative to plant life form

experience and lessons in the protection and reintroduction of rare and endangered plants, and launch the next Five-Year Plan for plant protection and reintroduction. (3) Increase investment in plant protection and reintroduction. (4) Strengthen scientific research and formulate technical standards for protection and reintroduction and maximum genetic diversity in ex situ collections and reintroductions and sustainability use in agriculture, horticulture, and forestry. (5) The effect of plant protection should be included in the performance evaluation of local leaders, and the quality of plant protection managers should be improved through training. (6) Increase publicity and raise public awareness of plant protection.

These problems and recommendations of China also have reference value for other countries in the world.

References

J. Akeroyd, J.P. Wyse, *A Handbook for Botanic Gardens on the Reintroduction of Plants to the Wild* (Richmond, Botanic Gardens Conservation International, 1995)

M.A. Albrecht, E.O. Guerrant, J. Maschinski, et al., A long-term view of rare plant reintroduction. Biol. Conserv. **144**, 2557–2558 (2011)

G.M. Ames, D.A. Wall, M.G. Hohmann, et al., Functional trait similarity predicts survival in rare plant reintroductions. Ecol. Appl. e02087 (2020). https://doi.org/10.1002/eap.2087

D.P. Armstrong, P.J. Seddon, Directions in reintroduction biology. Trends Ecol. Evol. **23**, 20–25 (2008)

L.E. Commander, D. Coates, L. Broadhurst, *Guidelines for the Translocation of Threatened Plants in Australia*, 3rd edn. (Canberra, Australian Network for Plant Conservation, 2018)

Council of Europe, Recommendation No. R. (85)15 of the Committee of Ministers on the reintroduction of wildlife species. 4th Meeting; 1985

J. G. Ewen, D. P. Armstrong, K. A. Parker, P. J. Seddon (eds.), *Reintroduction Biology: Integrating Science and Management* (Wiley, Chichester, 2012)

D.A. Falk, C.J. Millar, M. Olwell, *Restoring Diversity: Strategies for Reintroduction of Endangered Plants* (Island Press, Washington, 1996)

S. Godefroid, C. Piazza, G. Rossi, et al., How successful are plant species reintroductions? Biol. Conserv. **144**, 672–682 (2011)

B. Griffith, J.M. Scott, J.W. Carpenter, et al., Translocation as a species conservation tool: status and strategy. Science **245**, 477–480 (1989)

V. Heywood, K. Shaw, Y. Harvey-Brown, et al., *BGCI and IABG's Species Recovery Manual* (Botanic Gardens Conservation International, Richmond, 2018)

H.W. Huang, Z. Zhang, The status and prospect of plant introduction and *ex situ* conservation in China. Biodiversitas **20**, 559–571 (2012)

IUCN, *Guidelines for Re-introductions. Prepared by the IUCN/SSC Re-introduction Specialist Group* (IUCN, Gland, 1998)

IUCN, *Guidelines for the in situ Re-introduction and Translocation of African and Asian Rhinoceros* (IUCN, Gland, 2009)

Li N, Long DD, Wang YH, et al. *Guidelines on Reintroduction Practice of Rare and Endangered Plants*. Forestry Industry Standards of the People's Republic of China, LY/T 2589 (2016)

G.H. Liu, J. Zhou, D.S. Huang, et al., Spatial and temporal dynamics of a restored population of Oryza rufipogon in Huli Marsh, South China. Restor. Ecol. **12**, 456–463 (2004)

H. Liu, C.L. Feng, B.S. Chen, et al., Overcoming extreme weather events: successful but variable assisted translocations of wild orchids in southwestern China. Biol. Conserv. **150**, 68–75 (2012)

H. Liu, H. Ren, Q. Liu, et al., The conservation translocation of threatened plants as a conservation measure in China: a review. Conserv. Biol. **29**, 1537–1551 (2015)

Y.P. Ma, G. Chen, R.E. Grumbine, et al., Conserving plant species with extremely small populations (PSESP) in China. Biodivers. Conserv. **22**, 803–809 (2013)

J. Maschinski, K.E. Haskins, *Plant Reintroduction in a Changing Climate: Promises and Perils* (Island Press, Washington, 2012)

M. Maunder, Plant reintroduction: an overview. Biodivers. Conserv. **1**, 51–61 (1992)

T. Polak, D. Saltz, Reintroduction as an ecosystem restoration technique. Conserv. Biol. **25**, 424–425 (2011)

H. Qin, Y. Yang, S. Dong, et al., Threatened Species List of China's Higher Plants. Biodivers. Sci. **25**(7), 696–744 (2017)

H. Ren, Z.R. Duan, *The Theory and Practice of Building Scientific Botanical Gardens*, 2nd edn. (Science Press, Beijing, 2017)

H. Ren, G. Ma, Q. Zhang, et al., Moss is a key nurse plant for reintroduction of the endangered herb, Primulina tabacum Hance. Plant Ecol. **209**, 313–320 (2010)

H. Ren, Q.M. Zhang, H.F. Lu, et al., Wild plant species with extremely small populations require conservation and reintroduction in China. Ambio **41**, 913–917 (2012a)

H. Ren, S.J. Zeng, L. Li, Community ecology and reintroduction of *Tigridiopalma magnifica*, a rare and endangered herb. Oryx **46**, 391–398 (2012b)

H. Ren, S.G. Jian, H.X. Liu, et al., Advances in the reintroduction of rare and endangered wild plant species. Sc. China Life Sci. **57**, 603–609 (2014)

H. Ren, J. Wang, H. Liu, et al., Conservation introduction resulted in similar reproductive success of *Camellia changii* compared with augmentation. Plant Ecol. **217**, 219–228 (2016a)

H. Ren, H. Liu, J. Wang, et al., The use of grafted seedlings increases the success of conservation translocations of *Manglietia longipedunculata* (Magnoliaceae), a critically endangered tree. Oryx **50**, 437–445 (2016b)

H. Ren, Q. Liu, L.H. Li, et al., *Introduction to Restoration Ecology*, 3rd edn. (Science Press, Beijing, 2019)

P.J. Seddon, From reintroduction to assisted colonization: moving along the conservation translocation spectrum. Restor. Ecol. **18**, 796–802 (2010)
P.J. Seddon, A. Moehrenschlager, J. Ewen, Reintroducing resurrected species: selecting DeExtinction candidates. Trends Ecol. Evol. **29**, 140–147 (2014)
V.A. Sheean, A.D. Manning, D.B. Lindenmayer, et al., An assessment of scientific approaches towards species relocations in Australia. Austral. Ecol. **37**, 204–215 (2012)
J.L. Silcock, Aboriginal translocations: the intentional propagation and dispersal of plants in Aboriginal Australia. J. Ethnobiol. **38**, 372–405 (2018)
G. Taylor, S. Canessa, R.H. Clarke, et al., Is reintroduction biology an effective applied science? Trends Ecol. Evol. **32**, 873–880 (2017)
Q.G. Ye, S.S. Chen, Clarification of indigenous and exotic rare and endangered plant species in Jiugongshan Nature Reserve. J. Wuhan Bot. Res. **22**, 78–81 (2004)
Q.G. Ye, S.S. Chen, S.Y. Wang, Construction of communities for *ex situ* conservation of the rare and endangered plants in the Three Gorge Reservoir area, in *Biodiversity Conservation and Regional Sustainable Development: 4th National Biodiversity Conservation and Sustainable Utilization Symposium Collected Papers*, ed. by Y. Y. Chen, K. P. Ma, (China Forestry Publishing, Beijing, 2002), pp. 162–168
X. Zhou, J.Y. Gao, Reintroduction of rare and endangered plant species: theory and practice. Biodiversity. **19**, 97–105 (2011)

Part II
Reintroduction Cases of Rare and Endangered Plants in China

Reintroduction of *Primulina tabacum* Hance, a Critically Endangered Calciphilous Perennial Herb, in Southern China

Hai Ren, Guohua Ma, Qianmei Zhang, and Xiangying Wen

Abstract *Primulina tabacum* Hance (Gesneriaceae) is a calciphilous perennial herb. There are only eight wild populations with less than 10,000 individuals in Southern China. The distribution, conservation status, ecological and biological characteristics, genetic diversity, reproductive biology, tissue culture, and horticulture of *P. tabacum* were studied before reintroduction. Thousand in vitro-propagated *P. tabacum* plantlets were used to reintroduce at three of the plant's historical and extant habitats. About 10% of the transplanted seedlings survived by 2012. More than 200 next-generation individuals were found in 2018. Facilitation between the species and mosses is important for the reintroduction success.

Keywords Calciphilous perennial herb · Tissue culture · Moss · Facilitation · *Primulina tabacum* · Reintroduction

1 Introduction

Primulina tabacum Hance (Gesneriaceae) is a calciphilous perennial herb (Fig. 1). It was listed as one of the prioritized protected key wild plants of China in 1999 and in the list of wild plants with extremely small populations in China in 2012 (Ren et al. 2010a). It only distributes at the entrances of karst cave drainages along the border among northern Guangdong, southern Hunan, and eastern Guangxi, China. There are only eight wild populations at the region. *P. tabacum* relies on alkaline calciferous groundwater and grows in poor soils (Fig. 2). Because of human disturbances and climate change, the population size of *P. tabacum* has drastically decreased during the past century. It was estimated that there were less than 10,000 individuals in the wild.

H. Ren (✉) · G. Ma · Q. Zhang · X. Wen
South China Botanical Garden, Chinese Academy of Sciences, Guangzhou, China
e-mail: renhai@scbg.ac.cn; magh@scbg.ac.cn; zqm@scbg.ac.c; wx-ying@scib.ac.cn

H. Ren (ed.), *Conservation and Reintroduction of Rare and Endangered Plants in China*, https://doi.org/10.1007/978-981-15-5301-1_5

Fig. 1 The individual of *Primulina tabacum* Hance

Fig. 2 The habitat of *Primulina tabacum* Hance

2 Description of Reintroduction

2.1 *Feasibility*

We have studied the distribution, conservation status, ecological and biological characteristics, genetic diversity (Wang et al. 2013), reproductive biology, tissue culture, and horticulture of *P. tabacum* since 2002 (Ren et al. 2010a). We established reintroduction sites at Lianzhou City, Guangdong, Southern China in 2002. We failed to germinate the seeds at the South China Botanical Garden in 2003–2006. However, we got successful in tissue culture and obtained plantlets in 2007. We used in vitro-propagated *P. tabacum* plantlets to reintroduce the species into three of the plant's historical and extant habitats (Ren et al. 2010b).

2.2 *Implementation*

We obtained about 4000 plantlets in July and acclimatized these plantlets at the South China Botanical Garden until September 2007. During the acclimation period, 7.2% of the plantlets died from desiccation. Of the remaining plantlets, 1000 were then transplanted into the cave entrances at Dixiahe (112°21′ E, 25°1′ N), Lianzhou City, Guangdong, Southern China, on October 26, 2007. At the time of transplanting, the plantlets were 1.5 ± 0.1 cm in height and 3.0 ± 1.0 cm $\times$ 3.5 ± 1.0 cm in crown size. The transplants were kept natural growth except for watering several times after transplanting. The planting plots were not fenced nor fertilized. In addition, we proposed successfully to the local government to establish a small natural reserve to protect the remaining wild individuals in 2007. We also successfully established an ex situ collection in the experimental area of the Tianxin nature reserve, Lianzhou City, in 2010 (Ren et al. 2010b, 2018).

2.3 *Post-planting Monitoring*

After transplantation, we monitored the survival, height, and crown of all transplants and examined the causes of death (i.e., insect defoliation, fungal decay, nutrient deficiency, lack of water, or strong radiation) from 2007 to 2012. The monitoring was carried out once per month during the first year and once per year thereafter (Fig. 3.). Microhabitats and soils were also monitored every year. About 10% of the transplanted seedlings survived by 2012. We also found more than 200 next-generation individuals in 2018. Our field observations indicate that transplanted *P. tabacum* grew slower than wild *P. tabacum*. The transplanted *P. tabacum* performed especially well under the cover of the nursing moss, *Gymnostomiella*

Fig. 3 Monitoring the growth of *Primulina tabacum* reintroduced individuals

longinervis Broth. Facilitation between the species and mosses is important for the reintroduction success (Ren et al. 2010b, 2018).

3 Problems and Recommendations

- There may be some remaining populations in remote mountain areas; thus more surveys are needed.
- The local farmers and domestic animals unintentionally disturbed or sometimes destroyed the reintroduction sites.
- Successful reintroduction needs the close collaboration among all stakeholders, including farmers, scientists, and the local government officials.
- Moss plays a key nurse plant to facilitate the reintroduction of *P. tabacum*.
- It is easy to succeed in reintroduction of rare and endangered plants by the integration of biotechnology, nurse plant technology, and ecological restoration technology.
- The best method for conservation of rare and endangered plant is in situ preservation, and reintroduction can be used as a helpful tool to conserve biodiversity, but it is difficult and expensive.

Acknowledgments This chapter was modified from the paper published by our group in *Global Re-introduction Perspectives: 2018* (Ren et al. 2018; 243–246). The related contents are reused with the permission.

References

H. Ren, Q.M. Zhang, Z.F. Wang, et al., Conservation and possible reintroduction of an endangered plant based on an analysis of community ecology: A case study of *Primulina tabacum* Hance in China. Plant Species Biol. **25**, 43–50 (2010a)

H. Ren, G.H. Ma, Q.M. Zhang, et al., Moss is a key nurse plant for reintroduction of the endangered herb, *Primulina tabacum* Hance. Plant Ecol. **209**, 313–320 (2010b)

H. Ren, G.H. Ma, Q.M. Zhang, et al., The re-introduction of *Primulina tabacum* Hance, a critically endangered, endemic plant in southern China, in *Global Re-introduction Perspectives: 2018. Case Studies From Around the Globe*, ed. by P. S. Soorae, (IUCN/SSC Reintroduction Specialist Group and Abu Dhabi, UAE: Environment Agency-Abu Dhabi, Gland, 2018), pp. 243–246

Z.F. Wang, H. Ren, Z.C. Li, et al., Local genetic structure in the critically endangered, cave-associated perennial herb *Primulina tabacum* (Gesneriaceae). Biol. J. Linn. Soc. **109**, 747–756 (2013)

References

Artificial Propagation and Reintroduction of *Dayaoshania cotinifolia* W. T. Wang, an Extremely Small Population in China

Guohua Ma, Beiyi Guo, Songjun Zeng, Yubing Wang, Dehua Tu, Jinfeng Lü, and Hai Ren

Abstract *Dayaoshania cotinifolia* W. T. Wang is a perennial herb which is an endemic species in the Guangxi Zhuang Autonomous Region, China. It is a relatively primitive species of the Gesneriaceae. It was listed as one of the prioritized protected key wild plants of China and a critical endangered species because of its small population, narrow distribution, and disturbance of human activity. We collected some fruits of *D. cotinifolia* to South China Botanical Garden for seed germination test. It showed that gibberellin (GA_3) could enhance seed germination, which indicated that the seed has dormancy period after seeds become matured. In another test, we inoculated the sterilized seeds on MS medium for in vitro culture. Half a month later, the seed germinated and the initiate cotyledons were used as explants to induce adventitious shoots. After that, an efficient in vitro propagation and regeneration system via shoot organogenesis was established from young leaf explants. Plantlets were transplanted to a mixture of sand, vermiculite, and humus (1:1:1) with 92% survived. We have totally propagated 1000 relatively strong plantlets, and they were transplanted in the mountain slopes in May of 2013 in Jinxiu Yaoshan Nature Reserve. About 452 plants were still alive after 1 year as we returned to investigate the reintroduction area. One and half years later, about 126 plants had grown up and flowered and even seed reproduction was noted. In October 2017, over 500 plantlets from cutting propagation of leaves were

G. Ma (✉) · S. Zeng · J. Lü · H. Ren
South China Botanical Garden, Chinese Academy of Sciences, Guangzhou, China
e-mail: magh@scbg.ac.cn; zengsongjun@scib.ac.cn; renhai@scbg.ac.cn

B. Guo
South China Botanical Garden, Chinese Academy of Sciences, Guangzhou, China

University of Chinese Academy of Sciences, Chinese Academy of Sciences, Beijing, China

Y. Wang
China Three Gorges University, Yichang, Hubei, China

D. Tu
Laoshan Nature Reserve of Jinxiu in Guangxi Zhuang Nationality Autonomous Region, Jinxiu, Guangxi, China

H. Ren (ed.), *Conservation and Reintroduction of Rare and Endangered Plants in China*, https://doi.org/10.1007/978-981-15-5301-1_6

transplanted in Jinxiu Yaoshan Nature Reserve. Three hundred eighty plants were still alive after 1 year later, and most of them flowered and seed reproduction was also noted. These results indicated that reintroduction of *D. cotinifolia* was successful.

Keywords *Dayaoshania cotinifolia* W. T. Wang · Seed germination · In vitro propagation · Regeneration system · Reintroduction

1 Introduction

Dayaoshania cotinifolia W. T. Wang is a perennial herb which is an endemic species in the Guangxi Zhuang Autonomous Region, China. It is a relatively primitive species of the Gesneriaceae, with high scientific research and economical (ornamental and medicine) value (Fig. 1). According to Guangxi Rare Plant Survey, *D. cotinifolia* has a distribution area of about 0.2 hm^2, with a population of about 9600 individuals in 1999. It was listed as one of the prioritized protected key wild plants of China and a critical endangered species because of its small population, narrow distribution, and disturbance of human activity (Wang and Xie 2004). A comprehensive wild survey showed that the *D. cotinifolia* population has decreased significantly to about 3000 individuals during 2004 and 2006 and about 2,000 individuals by our comprehensive investigation in 2017 (Fig. 2). Up to now, there are very few reports on *D. cotinifolia*'s natural profile, distribution status, biological characteristics, and causes of endangerment (Wang et al. 2008, 2011). In order to preserve this rare and endangered plant and protect its resources, we carried out wild

Fig. 1 The individual of flowering *Dayaoshania cotinifolia* in Jinxiu, Guangxi

Fig. 2 Resource investigation of *Dayaoshania cotinifolia* in 2017

investigation, researched its geographical distribution and resource status, and put forward the research on artificial propagation via biotechnology and reintroduction during 2010–2019.

2 Description of Reintroduction

2.1 Feasibility

Jinxiu Yaoshan Nature Reserve is the only remaining distribution area of *D. cotinifolia*, located in the Guangxi Zhuang Autonomous Region (109°54′–110°15′E, 23°43′–24°09′N), with an area of 8875 hm^2, which is located in the middle to southwest of the Yaoshan Mountain. The reserve is a mountainous landscape with a minimum altitude of 300 m and a maximum altitude of 1643 m. The protection area is located in the transition zone of tropical Central Asia and South Asia, coupled with the complex terrain of the protected area, wild forest, with obvious mountain climate characterized by warm winter and cool summer, rainy days, less sunshine, and high humidity (Fig. 3). In the vegetation zoning, Jinxiu Yaoshan Nature Reserve belongs to the central part of China-Japan forest subregion in Southern China and the belt vegetation is the evergreen broadleaf forest in South Asia.

Fig. 3 The habitat of *Dayaoshania cotinifolia* in Jinxiu, Guangxi

Fig. 4 *Dayaoshania cotinifolia* seed germination treated with 200 mg/L GA_3 and plantlets grew up in sandy substrate within 3 months

2.2 Implementation

We collected some fruits of *D. cotinifolia* to South China Botanical Garden for seed germination test. It showed that GA_3 could enhance seed germination and grow up (Fig. 4), which indicated that the seed has dormancy period after seeds become matured. In another test, we inoculated the sterilized seeds on MS medium for in vitro culture. Half a month later, the seed germinated and the initiate cotyledons were used as explants to induce adventitious shoots. After that, an efficient in vitro propagation and regeneration system via shoot organogenesis was established from

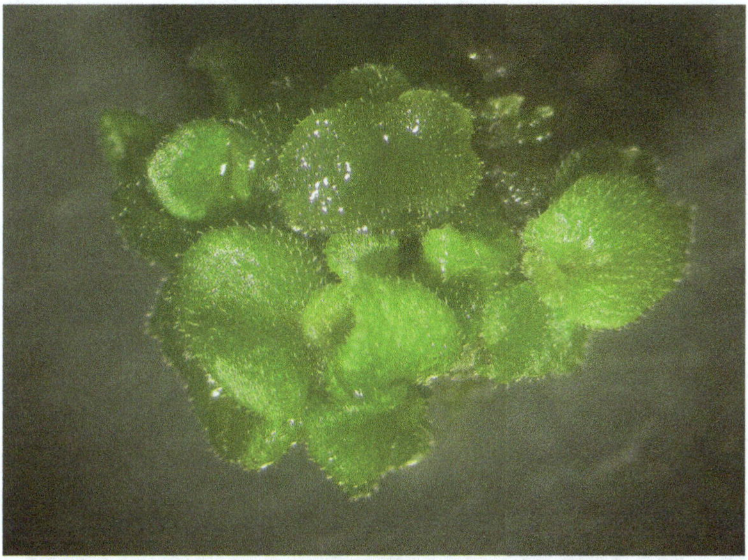

Fig. 5 Shoot organogenesis from leaf segment of *Dayaoshania cotinifolia* on the medium supplemented with 1.0 μM α-naphthalene acetic acid and 1.0–3.0 μM 6-benzyladenine

young leaf explants (Yang et al. 2014). Induction medium supplemented with 1.0 μM α-naphthalene acetic acid and 1.0–3.0 μM 6-benzyladenine was optimal for inducing adventitious shoots (Fig. 5). Induction medium containing 2.0 μM 6-benzyladenine and 0.5 μM indole-3-acetic acid was optimal for the multiplication of adventitious shoots. Rooting was achieved on half strength MS medium supplemented with 3.0 μM indole-3-acetic acid or α-naphthalene acetic acid and 0.1% activated charcoal. Plantlets were transplanted to a mixture of sand, vermiculite, and humus (1:1:1), 92% survived (Fig. 6).

2.3 *Post-planting Monitoring*

We have propagated about 2000 plantlets. Among them, 1000 stronger plantlets were transplanted in the mountain slopes in May of 2013 in Jinxiu Yaoshan Nature Reserve. One year later, as we returned to investigate the reintroduction area, about 452 plants were still alive (Fig. 7). One and half years later, about 126 plants had grown up and flowered and even seed reproduction was noted (Fig. 8). In October 2017, other 500 plantlets from cutting propagation of leaves were transplanted in Jinxiu Yaoshan Nature Reserve (Fig. 9), 380 plants were still alive after 1 year later, and most of them flowered and seed reproduction was also noted. These results indicated that reintroduction of *D. cotinifolia* was successful.

Fig. 6 In vitro plantlets of *Dayaoshania cotinifolia* were transplanted in a mixture of sand, vermiculite, and humus (1:1:1) for 1 month and 92% plantlets survived

Fig. 7 Reintroduction of *Dayaoshania cotinifolia* in 2017

Fig. 8 Reintroduction scene of *Dayaoshania cotinifolia* in Jinxiu, Guangxi, in 2014

Fig. 9 After reintroducing in habitat for one and half years, the plant *Dayaoshania cotinifolia* begins to flower in 2015

3 Problems and Recommendations

The plants of *D. cotinifolia* in the mountains are small and grow in a shady and humid environment. The individuals of *D. cotinifolia* growing on the rocks under the forest cover (Fig. 2) are usually stronger than those growing on the ground with direct sunshine, which may be related to the humid environment in the mountains, the shallow roots, the changes in the air humidity, and soil layer to which seedlings are more sensitive. *D. cotinifolia* is only suitable for survival in the scattered light environment, strictly dependent on the shady forest environment.

The fruit of *D. cotinifolia* is in a linear shape, with 1.9–5.5 cm long, and contents about 1000–2000 seeds. The seeds are very small, long, and oval, and 1000 seeds weigh approximately 16 mg. Due to the small size of the seed, the seed is difficult to track after being released from the fruit, but the seed germination rate in the mountains is low. This may be related to the pollination pathway in Yaoshan Mountain, which is in the under-forest herb layer and has not yet been found to be pollinated by some insects. However, it is worth noting that the pests and diseases of the fruit on Yaoshan Mountain are quite serious. Once the seed becomes matured, it overcomes a cold winter to germinate (the lowest temperatures in the area reach −5 °C), which seriously affects seed viability and seed germination rate. The wild survey rarely found visible seedlings of *D. cotinifolia* in the large population of this endangered plant species in Yaoshan Mountain.

References

S. Wang, Y. Xie, *Chinese Species Red List*, vol 1 (Higher Education Press, Beijing, 2004)

Y.B. Wang, H.W. Liang, F.J. Chen, et al., The endangered causes and protecting strategies for *Dayaoshania cotinifolia*, an endemic plant in Guangxi. Ecol. Environ. **17**, 1956–1960 (2008)

Y.B. Wang, H.W. Liang, N.B. Mo, et al., Flower phenology and breeding system of rare and endangered *Dayaoshania cotinifolia*. Acta Bot. Boreal Occident Sin. **31**, 861–867 (2011)

G. Yang, J.F. Lü, J.A. Teixeira da Silva, et al., Shoot organogenesis from leaf explants of *Dayaoshania cotinifolia* W. T. Wang. In Vitro Cell Dev. Biol. Plant **50**, 451–457 (2014)

Reintroduction of *Tigridiopalma magnifica* C. Chen, a Critically Endangered Endemic Plant in China

Songjun Zeng, Hai Ren, Longna Li, and Kunlin Wu

Abstract *Tigridiopalma magnifica* is a rare and endemic perennial herb plant, the only species in the genus *Tigridiopalma* in in the Melastomatacea family in Yangchun, Xinyi, and Gaozhou cities of China, where it is categorized as Critically Endangered on the national Red List. Twelve locations with less 10,000 individuals of *T. magnifica* have been identified (1 extinct, 11 extants). *T. magnifica* only grows in the surface soil on stone walls or rocks under the canopy of secondary forests and plantations and has no specific associated plant species. An efficient propagation in vitro system using seeds and leaves as explants has been established for *T. magnifica* C. Chen. About 20,000 well-developed root systems were transferred to planting pots containing a mixture of sand, sieved peat, and perlite (1:1:1; v/v) in greenhouse conditions. More than 80.0% of the plantlets survived after 30 days. We reintroduced 10,000 plantlets produced by tissue culture into three reintroduction locations in and outside the current and historical distribution area of this species. After 11 months survival rate was 40–58%, while the survival was higher and plantlet crowns were larger at the location within the original range of the species than at the other two sites.

Keywords *Tigridiopalma magnifica* · Habitat · In vitro propagation · Regeneration system · Reintroduction

S. Zeng (✉) · H. Ren · K. Wu
South China Botanical Garden, Chinese Academy of Sciences, Guangzhou, China
e-mail: zengsongjun@scib.ac.cn; renhai@scbg.ac.cn

L. Li
South China Botanical Garden, Chinese Academy of Sciences, Guangzhou, China

Laboratory Center of Life Sciences, Nanjing Agricultural University, Nanjing, China

H. Ren (ed.), *Conservation and Reintroduction of Rare and Endangered Plants in China*, https://doi.org/10.1007/978-981-15-5301-1_7

1 Introduction

Tigridiopalma magnifica C. Chen is a rare and endemic perennial herb plant in the Melastomatacea family in Yangchun, Xinyi, and Gaozhou cities of China (Fig. 1), where it is categorized as Critically Endangered on the national Red List (Chen 1979; Zhang and Miao 1984). This herb also has the potential as an ornamental plant because it is tolerant of shade and has charming leaves with reddish villi and beautiful flowers (Zeng 2005). Because of habitat destruction, human disturbances, and climate change, the population size of *T. magnifica* has drastically decreased during the past century. Twelve locations with populations of *T. magnifica* have been identified (1 extinct, 11 extants). *T. magnifica* only grows in the surface soil on stone walls or rocks under the canopy of forests and plantations and has no specific associated plant species. Canopy closure, soil water content, and the distance to the closest stream are the main factors influencing the distribution of *T. magnifica* (Fig. 2). It was estimated that there were less than 10,000 individuals of this endangered plant in the wild (Ren et al. 2012).

Fig. 1 The individual of *Tigridiopalma magnifica* C. Chen

Fig. 2 The habitat of *Tigridiopalma magnifica* C. Chen

2 Description of Reintroduction

2.1 Feasibility

We studied the distribution (Fig. 3), conservation status, ecological and biological characteristics, reproductive biology, tissue culture (Zeng et al. 2008), and horticulture of *T. magnifica* since 2003 (Zeng 2005; Ren et al. 2012). An efficient propagation in vitro system using seeds and leaves as explants has been established for *T. magnifica* C. Chen, and about 20,000 well-developed root systems were transferred to planting pots containing a mixture of sand, sieved peat, and perlite (1:1:1; v/v) in greenhouse conditions. Over 80.0% of the plantlets survived after 30 days, and the regenerated plantlets could flower after 2-year transplanting (Fig. 4). We established three reintroduction sites at Dajiangkou Village (111°30′ E, 21°55′ N) in Yangchun city, Ehuangzhang Village (111°46′ E, 22°20′ N) in Yangchun city, and

Fig. 3 Resource investigation of *Tigridiopalma magnifica* C. Chen

Fig. 4 Plant regeneration from seed explants of *Tigridiopalma magnifica*

Tianxin Forestry Farm (112°21′ E, 25°6′ N) in Lianzhou city, Guangdong, Southern China in 2008 (Fig. 5) with the regenerated plantlets transplanted for 1 or 2 years. The first location is an augmentation reintroduction within an extant population. The second location is a reintroduction site within the historical range of *T. magnifica* but

Fig. 5 Reintroduction of *Tigridiopalma magnifica* C. Chen

had no population at present. The third location, Tianxing Forestry Farm, is an introduction site outside the historical range of the species. The structure of the plant community at the third site is similar to that at Ehuangzhang Village (Ren et al. 2012).

2.2 *Implementation*

We obtained about 20,000 plantlets by propagation in vitro and acclimatized these plantlets at the South China Botanical Garden in 2007. Among them, 10,000 plantlets were used in reintroduction. During the reintroduction acclimation period, the plantlets were watered during the first 3 days after transplantation but not thereafter. The quadrats were not fenced, fertilized, or mulched. In addition, we proposed successfully to the local government to establish a small natural reserve to conserve the remaining wild and reintroduced individuals of this endangered plant in 2010. Thirty-four months after transplantation, the survival rate was 45%, 33%, and 33% at Dajiangkou, Ehuangzhang, and Tianxin, respectively. We also successfully established an ex situ collection in the experimental area of the Tianxin nature reserve, Lianzhou city, in 2010 (Ren et al. 2012).

Fig. 6 Flowering of plant reintroduced by *Tigridiopalma magnifica* C. Chen

2.3 *Post-planting Monitoring*

After transplanting, we observed and monitored the survival, height, and crown of all transplants from 2010 to 2017. The monitoring was carried out once per 3 months during the first year and once per year thereafter. About 60% of the transplanted plantlets survived by 2016, and possible causes of some transplants' mortality were defoliation by the larvae of a noctuid moth, strong radiation, and lack of water. We found that transplanted plantlets grew slower than wild *T. magnifica*. Although all the surviving individuals had flowered and produced seeds by 34 months at the locations where they were reintroduced, the germination rate in the field was low because seed germination needs illumination, while the seeds are tiny and are easy to be washed into the soil without illumination. However, we also found more than 500 next-generation individuals in 2017 (Fig. 6).

3 Problems and Recommendations

- It is essential to have an integrated plan for species recovery of *T. magnifica*, which includes thorough investigation of germplasm resources, in situ and ex situ conservation, establishment of efficient propagation systems, and prevention of plant removal.
- There may be some remaining populations at other locations in this area, which need further investigating. The genetic diversity of all surviving *T. magnifica* populations should be analyzed by molecular tools as a support to the protection to all germplasm resources with different genetic diversities.

- An efficient propagation system is necessary to produce lots of plantlets for commercial use, which can reduce the demand for wild resources.
- The best method for conservation of rare and endangered plant is in situ preservation, Reintroduction can be used as a helpful tool to conserve biodiversity, but it is difficult and expensive. However, the local farmers unintentionally disturbed or sometimes destroyed the reintroduction sites, which can only be resolved by legislation, public education, and suitable economic compensation.

References

J. Chen, A new genus *Tigridiopalma* C. Chen of Melastomataceae. Acta Bot. Yun. **1**, 106–109 (1979)

H. Ren, S.J. Zeng, L.N. Li, et al., Community ecology and reintroduction of *Tigridiopalma magnifica*, a rare and endangered herb. Oryx **46**(3), 371–398 (2012)

S.J. Zeng, Rare foliage plants—*Tigridiopalma magnifica*. Flower Pot Garden. **12**, 39 (2005)

S.J. Zeng, J. Duan, L.N. Li, et al., Plant regeneration from leaf explants of *Tigridiopalma magnifica* (Melastomataceae). Pak. J. Bot. **40**, 1179–1184 (2008)

H.D. Zhang, Y.H. Miao, *Flora of China* (Science Press, Beijing, 1984)

Reintroduction of *Camellia changii* Ye (Theaceae), a Critically Endangered Plant Endemic to Southern China

Hai Ren, Hongxiao Liu, Qianmei Zhang, Yi Xu, Ju Zhou, and Yan Zeng

Abstract *Camellia changii* Ye is an evergreen tall shrub or small tree with an extremely narrow distribution. There are only 1039 individuals in the field. The community characteristics, genetic diversity, and reproductive barriers of *C. changii* were studied. An ex situ living collection was established, which contained the entire wild genetic diversity by grafting techniques. Grafted plants were used to conduct augmentation and conservation introduction at two sites. All transplants grew successfully after 2 years at the two sites. We concluded that the out-of-range conservation introduction of *C. changii* did not reduce reproductive success compared with augmentation.

Keywords Global change · Conservation introduction · Grafting plant · Augmentation · Genetic diversity · Reintroduction

1 Introduction

Camellia changii Ye is an evergreen tall shrub or small tree in South China. It is an insect-pollinated and self-incompatible species. It has 1039 individuals in the sole wild population. *C. changii* is endemic to an extremely narrow geographical scope. The species was documented by the conservation list of extremely small populations in China in 2012 (Ren et al. 2014). It is known among Chinese botanists as the "giant

H. Ren (✉) · H. Liu · Q. Zhang
South China Botanical Garden, Chinese Academy of Sciences, Guangzhou, China
e-mail: renhai@scbg.ac.cn; liuhx@scbg.ac.cn; zqm@scbg.ac.cn

Y. Xu
Administration Ehuangzhang Nature Reserve, Yangchun, Guangdong, China

J. Zhou · Y. Zeng
Bureau of Science and Technology for Development, Chinese Academy of Sciences, Beijing, China
e-mail: zhouju@cashq.ac.cn; zengyan@cashq.ac.cn

H. Ren (ed.), *Conservation and Reintroduction of Rare and Endangered Plants in China*, https://doi.org/10.1007/978-981-15-5301-1_8

Fig. 1 The *Camellia changii* Ye in flower season

panda of the plant kingdom" because of its rarity and charm (Fig. 1, Wang and Xie 2004). *C. changii* were located within Ehuangzhang Nature Reserve in Yangchun County, Yangjiang City, south-west Guangdong Province (Fig. 2, Wu 1998; Zhang 1999). This wild population suffers to not only natural threats, such as global climate change, but also human collection due to its high ornamental value (Ren et al. 2014). Other threats include development of local tourism and recreation within the nature reserve, construction of small dams, and smallholder farming and also its restricted range, limited recruitment, and potential inbreeding depression as a result of the small population size and low genetic diversity (Luo et al. 2007).

2 Description of Reintroduction

2.1 Feasibility

Before our conservation practices, researches were conducted for *C. changii* concerning its community characteristics, genetic diversity, and reproductive barriers (Luo et al. 2007, 2008, 2011). Building on previous knowledge, we examined all specimens of *C. changii* deposited in the herbaria of the South China Botanical Garden, Chinese Academy of Sciences, and Sun Yat-sen University (Ren et al. 2014). Both herbaria are major depositories for collections of plant material in Southern China. We then surveyed for *C. changii* in the Ehuangzhang Nature Reserve from January 2009 to January 2013.

Fig. 2 The habitat of *Camellia changii* Ye

2.2 Implementation

From February 2013 to December 2014, we conducted experiment to compare the performance of two types of conservation translocation, i.e., augmentation and conservation introduction at two sites (Ehuangzhang and Tianxin, respectively; Ren et al. 2016). Attesting experiment, both the grafted and cutting plants survived and grew well, while the grafted plants performed a relatively higher survival rate and growth speed at both sites. Therefore, grafted plants were taken as materials. In March 2009, we grafted 300 *C. changii* scions (2–3 cm each) to 300 rootstocks of 1-year-old *Camellia gauchowensis*. Of the 300 grafted plants, more than 280 grew wells and began to bloom in 2012. In January 2013, we planted 45 grafted plants at both Ehuangzhang and Tianxin site. A 1 ha experimental field was set in each site. Three plots were partied in each experimental field. Each plot was future divided into a 3 × 3 m grid systems and 15 grafted plants were planted. We irrigated the plants several times after transplanting. Since *C. changii* plants are sun-tolerant, we removed all trees and shrubs from all plots before transplantation to avoid competition. Afterward no management measures were taken to these plants. The plots were not fenced, fertilized, nor mulched (Ren et al. 2016).

2.3 Post-planting Monitoring

The survival rate and height of each plant were measured every 6 months for 2 years from January 2013 to December 2014 (Fig. 3). Floral phenology, pollinator, and fruit and seed productions were observed and anthocyanin content of petals was analyzed. We recorded the flowers from each plant every week during 2013 and 2014. Flower visitors were observed at both sites during the peak flowering period for 12 days (3 days in May, July, September, and October, respectively) each year. In each plot, we harvested the petals of five flowers from individual *C. changii* and analyzed the anthocyanin content of these petals (Ren et al. 2016). All of the *C. changii* were successfully transplanted after 2 years at both sites. Detailed result can be found in Ren et al. (2016). We concluded that out-of-range conservation introduction of *C. changii* did not have a lower reproductive success than augmentation.

Fig. 3 The transplanted seedling of *Camellia changii* Ye

3 Problems and Recommendations

- An integrative species conservation plan for the species that includes patrolling the Ehuangzhang Nature Reserve to prevent plant removal may be beneficial.
- Other potential strategies include establishing an ex situ living collection that contains the entire wild genetic diversity (accomplished by grafting of short cuttings from all wild individuals), facilitating propagation for commercial use, and implementing reintroduction to augment the wild population.
- *C. changii* should be categorized as Critically Endangered on the IUCN Red List.

References

X.Y. Luo, X.Y. Zhuang, Y.S. Yang, Genetic diversity of *Camellia changii* Ye (Theaceae) using ISSR markers. J. Trop. Subtrop. Bot. **15**, 93–100 (2007)

X.Y. Luo, L.J. Mo, G.D. Tang, et al., The community characteristics of China's endangered and endemic species *Camellia changii*. J. Fujian Forest Sci. Technol. **35**, 63–68 (2008)

X.Y. Luo, G.D. Tang, L.J. Mo, et al., Pollination biology of *Camellia changii*. Chin. J. Ecol. **30**, 552–557 (2011)

H. Ren, S.G. Jian, Y. Chen, et al., Distribution, status, and conservation of *Camellia changii* Ye (Theaceae), a critically endangered, endemic plant in southern China. Oryx **48**, 358–360 (2014)

H. Ren, J. Wang, H. Liu, et al., Conservation introduction resulted in similar reproductive success of *Camellia changii* compared with augmentation. Plant Ecol. **217**, 219–228 (2016)

S. Wang, Y. Xie, *Chinese Species Red List*, vol 1 (Higher Education Press, Beijing, 2004)

T.L. Wu, *Flora of Guangdong*, vol 6 (Guangdong Science and Technology Press, Guangzhou, 1998)

H.T. Zhang, *Flora of China*, vol 49 (Science Press, Beijing, 1999)

Conservation Translocation of the Rare and Endangered Tree *Euryodendron excelsum* in South China

Hai Ren, Qianmei Zhang, Yiming Fan, Zeyuan Zou, and Yi Xu

Abstract *Euryodendron excelsum* H. T. Chang is an evergreen tree and a wild plant with extremely small populations. There are no more than 200 individuals of this species at ten isolated sites in the wild. The distribution, conservation status, ecological and biological characteristics, genetic diversity, reproductive biology and technique, seed biology, and cultivation of *E. excelsum* were studied. Reintroduction experiments including augmentation and translocation were conducted at Yangchun City and Guangzhou City of Guangdong, Hekou County, and Jinghong City of Yunna. The survival rate of cuttings was about 20%. Both the survival rate and growing speed of the augmentation individuals were greater than that of the translocated individuals. As a species with high level of genetic variation, rational collocation of subpopulation individuals of *E. excelsum* during reintroduction is needed.

Keywords Augmentation · Genetic diversity · Wild plant with extremely small population · Cutting

---

H. Ren (✉) · Q. Zhang
South China Botanical Garden, Chinese Academy of Sciences, Guangzhou, China
e-mail: renhai@scbg.ac.cn; zqm@scbg.ac.cn

Y. Fan
Foshan Institute of Science and Technology, Foshan, China

Z. Zou
South China Agricultural University, Guangzhou, China

Y. Xu
Administration Ehuangzhang Nature Reserve, Yangchun, Guangdong, China

H. Ren (ed.), *Conservation and Reintroduction of Rare and Endangered Plants in China*, https://doi.org/10.1007/978-981-15-5301-1_9

1 Introduction

Euryodendron excelsum H. T. Chang is an evergreen tree in the *Euryodendron* genus of family Theaceae (Fig. 1). It has been listed as a critically endangered species by the International Union for Conservation of Nature (IUCN) and, as a wild plant with extremely small populations, is considered as a second-class protected plant in China (Shen et al. 2009). *E. excelsum* generally grows under forests with latosol or lateritic red soil. There are no more than 200 individuals of this species in the wild (Fig. 2). They distribute at ten isolated sites in Yangchun City, Guangdong Province, South China. Individuals of *E. excelsum* show a high level of genetic variation both within and between populations because of habitat fragmentation (Ren et al. 2019). Two populations in neighboring Guangxi Zhuang Autonomous Region have become extinct because of human destruction (Shen et al. 2007).

Fig. 1 The individual, flowers, and fruit of *Euryodendron excelsum*

Fig. 2 The habitat of *Euryodendron excelsum*

2 Description of Reintroduction

2.1 Feasibility

Professor Wang Yuehua at Yunnan University and Professor Ren Hai at South China Botanical Garden of Chinese Academy of Sciences have studied the distribution, conservation status, ecological and biological characteristics, genetic diversity, reproductive biology and technique, seed biology, and cultivation of *E. excelsum* since 2000 (Shen et al. 2016; Ren et al. 2019). They established reintroduction sites at Yangchun City and Guangzhou City of Guangdong, Hekou County, and Jinghong City of Yunnan after 2002. *E. excelsum* seeds are nondormant, and about 70% germinate. The survival rate of cuttings is about 20% (Shen et al. 2016). Zhang (2018) reported a successful tissue culture for this species.

2.2 Implementation

They obtained about 400 seedlings by seed germination and conducted two experimental translocations. In one case, they augmented an existing population (Ehuangzhang Nature Reserve, Yangchun City). In the other case, they conducted conservation introduction, i.e., they introduced the species at a site outside of its known historical range (Hekou County of Yunnan Province; Jinghong City of Yunnan; South China Botanical Garden, Guangzhou City). The initial height of all 3-year-old seedlings was 35.1 ± 2.1 cm at the time of transplantation. The seedlings were watered and fertilized during the first month after transplanting. The experimental sites were not fenced. In addition, they proposed successfully to the local government to establish small protected sites for conserving the remaining wild individuals in 2012. They also established an ex situ conservation section at Ehuanzhang Nature Reserve in 2012 (Zhang 2014).

2.3 Post-planting Monitoring

After transplantation, they monitored the survival and height of all individuals (Fig. 3). The monitoring was carried out once per month during the first year and once per year thereafter. Microhabitats and soils were also monitored. About 90% of the seedlings survived after 38 months. The height of the augment individuals was

Fig. 3 Monitoring the growth of *Euryodendron excelsum* conservation translocation individuals

85.2 ± 5.1 cm, and that of the conservation introduced individuals was 56.5 ± 8.3 cm after 38 months. Both the survival rate and growing speed of the augmentation individuals were greater than that of the introduced individuals (Ren et al. 2019). The individuals at Ehuangzhang Nature Reserve grow up to 8 m and blossomed and yielded fruits after 2014.

3 Problems and Recommendations

- The local farmers or thieves occasionally intentionally stole the wild individuals.
- *E. excelsum* grows slowly, but arbuscular mycorrhizal fungi can increase seedling survival and growth.
- The endemic, rare, and endangered plants with narrow distributions may adapt to climate change by rapidly altering their morphological, anatomical, and physiological traits.
- As a species with high level of genetic variation, rational collocation of subpopulation individuals during reintroduction is needed.

References

H. Ren, H.L. Yi, Q.M. Zhang, et al., Morphological, anatomical and physiological traits of *Euryodendron excelsum* as affected by conservation translocation (augmentation vs. conservation introduction) in South China. Photosynthetica **57**, 217–225 (2019)

S.K. Shen, Y.H. Wang, Y.J. Zhang, et al., Advances research on the rare and endangered plant *Euryodendron excelsum* endemic to China. Chin. Wild Plant Res. **26**, 1–4 (2007)

S.K. Shen, Y.H. Wang, B.Y. Wang, et al., Distribution, stand characteristics and habitat of a critically endangered plant *Euryodendron excelsum* H. T. Chang (Theaceae): Implications for conservation. Plant Species Biol. **24**, 133–138 (2009)

S.K. Shen, F.Q. Wu, G.S. Yang, et al., Seed germination and seedling emergence of *Euryodendron excelsum* H. T. Chang: Implications for species conservation and restoration. Plant Species Biol. **31**, 233–239 (2016)

A. L. Zhang, Study on Reintroduction of Wild Plant with Extremely Small Population *Euryodendron excelsum* H. T. Chang, Doctoral Thesis at Yunnan University, Kunming, 2014

X. L. Zhang, Study on Micropropagation Technique of Rare and Wild Plant with Extremely Small Population, Master's Thesis at Zhongkai University of Agriculture and Engineering, Guangzhou, 2018

Conservation Translocations of *Manglietia longipedunculata*, a Critically Endangered Tree in South China

Hai Ren, Hongxiao Liu, Qianmei Zhang, Yan Zeng, Ju Zhou, Wenhui Zhu, and Xiaodong Cui

Abstract There are only 11 individuals of *Manglietia longipedunculata* in the wild. The species is with low reproductive success and cannot produce fruit under natural conditions. Augmentation and translocation of the species were conducted by emerged and grafted seedlings in 2009. More than 64% of seedlings of both augmentation and translocation grew well after 5 years' transplantation in both sites. Grafted seedlings had better growth rate and ecophysiological performance and reached flowering age earlier than the emerged ones. Therefore, grafted seedlings have superior adaptability to abiotic stress. Grafting may be a useful method for in situ augmentation and conservation translocations of this species and probably is also applicable to rare species having semblable reproductive disorders.

Keywords Grafting seedling · Cutting seedling · Augmentation · Reproductive obstacles

1 Introduction

Manglietia longipedunculata is an evergreen, canopy tree (Fig. 1). Currently this species has 11 individuals which all grow in evergreen broadleaf forest at the Nankunshan Nature Reserve, Longmen County, Guangdong Province, South China (Ren et al. 2016). *M. longipedunculata* is documented on the Red List of

H. Ren (✉) · H. Liu · Q. Zhang
South China Botanical Garden, Chinese Academy of Sciences, Guangzhou, China
e-mail: renhai@scbg.ac.cn; liuhx@scbg.ac.cn; zqm@scbg.ac.cn

Y. Zeng · J. Zhou
Bureau of Science and Technology for Development, Chinese Academy of Sciences, Beijing, China
e-mail: zengyan@cashq.ac.cn; zhouju@cashq.ac.cn

W. Zhu · X. Cui
Administration of Nankunshan Nature Reserve, Longmen, Guangdong, China

H. Ren (ed.), *Conservation and Reintroduction of Rare and Endangered Plants in China*, https://doi.org/10.1007/978-981-15-5301-1_10

Fig. 1 An individual of *Manglietia longipedunculata*

Magnoliaceae (Cicuzza et al. 2007). *M. longipedunculata* has large flowers, red fruit, and beautiful foliage and thus is a favorable decorative plant (Fig. 2). In addition, it can produce high-quality timber of high economic value (Fig. 1). The low reproductive success of *M. longipedunculata* can be attributed to protogyny, short period of stigma receptivity, and insufficient insect pollinators (Xie et al. 2011). It cannot produce fruit under natural conditions. Unfavorable climate and increase in temperature may contribute to the lack of pollinators to some extent. Currently, seeds can only be obtained via human aid pollination. However, 50% of the seeds are not viable (Fang et al. 2006).

2 Description of Reintroduction

2.1 Feasibility

Before our conservation effort, research was conducted on *M. longipedunculata*. Zeng and Law (2004) and Xie et al. (2011) had conducted a thorough survey in the Nankunshan Nature Reserve during 2007–2008. Xie et al. (2011) studied the causes

Fig. 2 The habitat of *Manglietia longipedunculata*

of the low reproductive success of this species, while Fang et al. (2006) studied the seed propagation of this family. Based on previous knowledge, we surveyed for existing wild *M. longipedunculata* during March 2010 and June 2013 and measured height and diameter of each located individual. Two mature resprouted branches from one *M. longipedunculata* stump at Nankunshan were manually pollinated in June 2008. Seeds were collected in September 2008, which provided material for conservation translocation.

2.2 Implementation

We implemented an augmentation at Nankunshan Nature Reserve and a conservation introduction at Tianxin Nature Reserve. The seeds from hand-pollination were planted instantly in the magnolia garden of the South China Botanical Garden.

About 1400 emerged seedlings grew successfully. In 2008, we grafted 200 *M. longipedunculata* scions (2–3 cm each) to 200 rootstocks of 1-year-old *Manglietia moto*. In March 2009, we randomly chose 145 emerged and 145 grafted healthy individuals of similar size, among which 100 seedings of each type were planted at Nankunshan. The other 45 individuals of each type were planted at Tianxin. At both Nankunshan and Tianxin site, a 1-ha experimental field was parted into three blocks. Each block has two 50 × 30 m plots. At Nankunshan site, we planted 33–34 emerged or grafted seedings in each plot. At Tianxin site, 15 seedlings of each type were planted in each plot. Seedlings were irrigated three times on the first, third, and seventh day after transplantation. All trees and shrubs were removed from all plots before transplantation to avoid competition. The plots were not fenced, fertilized, or mulched. No management measures were taken.

2.3 Post-planting Monitoring

In the augmentation and conservation introduction practice, we monitored the community ecology and ecophysiological indicators (Fig. 3). Plant height and diameter were measured instantly after transplantation and were monitored once a year for 5 years. If a transplant was dead, we examined possible causes including insect defoliation, fungal decay, nutrient deficiency, drought, or radiation stress, according to its appearance. We measured light-use and water-use efficiencies of these plants at Nankunshan on 3 and 4 June 2013 and at Tianxin on 12 and 13 June 2013. We also measured photosynthetic and physiological indicators including actual photochemical efficiency, photochemical fluorescence quenches, etc. (Ren et al. 2016). Environmental factors were collected at the same time. The seedlings become relatively stable after 1 year from transplantation. Five years' survive rate is over 64% at both sites and for both grafted and emerged seedlings. But grafted seedlings had better growth rate and ecophysiological performance than emerged ones. Grafted seedlings reach flower period earlier than the emerged seedlings. Therefore, grafted technique can be more adaptive to abiotic stress.

3 Problems and Recommendations

- Grafting technique could be useful in in situ augmentation and conservation translocations of *M. longipedunculata*. It is likely applicable to other rare species having alike reproductive disorders.
- A package conservation plan for this species was required including patrolling to prevent plant removal, establishment of an ex situ grafted living collection to keep genetic diversity, enhancing *M. longipedunculata* reproduction for commercial use, and conducting augmentation and translocation.

Fig. 3 Monitoring the eco-biological characteristics of *Manglietia longipedunculata*

- We suggest that hand-pollination can facilitate fruit set and thus aid the population spread both in situ and ex situ. More introductions should be conducted to augment existing populations.
- Tianxin may be more proper habitat for *M. longipedunculata* than Nankunshan, considering possible future temperature rise. This research implied that conservation introduction may be a feasible approach to sustain the species in the wild.

References

D. Cicuzza, A. Newton, S. Oldfield, *The Red List of Magnoliaceae* (Fauna & Flora International, Cambridge, 2007)

X.P. Fang, J. Xu, C.H. Yang, Propagation of seeds of ornamental plants in Magnoliaceae family. Seed **5**, 54–56 (2006)

H. Ren, H. Liu, J. Wang, et al., The use of grafted seedlings increases the success of conservation translocations of *Manglietia longipedunculata* (Magnoliaceae), a critically endangered tree. Oryx **50**(3), 437–445 (2016)

C. Xie, L. Fu, Q.W. Zeng, Studies on conservation of rare and endangered species *Manglietia longipedunculata* (Magnoliaceae), in *Proceedings of the 2nd International Symposium on the Family Magnoliaceae*, ed. by N. H. Xia, Q. W. Zeng, F. X. Xu, Q. G. Wu, (Huazhong Science and Technology University Press, Wuhan, 2011), pp. 149–162

Q.W. Zeng, Y.W. Law, *Manglietia longipedunculata* (Magnoliaceae), a new species from Guangdong, China. Ann Bot. Fennici. **41**, 151–154 (2004)

The Reintroduction of *Parakmeria omeiensis* Cheng, a Critically Endangered Endemic Plant, in Southwest China

Daoping Yu, Xiangying Wen, and Cehong Li

Abstract A total of 74 individuals of *Parakmeria omeiensis* in two small populations have been found in Mt. Emei, southwest China. This species is threatened by pollination difficulties, germination barriers, and habitat fragmentation. About 1750 saplings were propagated by seed germination and grafting, in which about 600 saplings were reintroduced to the wild in Mt. Emei with the altitude of 1200–1550 m. With the regular management and monitoring, 555 saplings are growing very well in two newly established populations, which increased the wild population numbers and sizes.

Keywords *Parakmeria omeiensis* · Magnoliaceae · Reintroduction · Extremely small populations · Critically endangered species

1 Introduction

Parakmeria omeiensis Cheng is an evergreen tree belonging to the family Magnoliaceae. It grows to 15–20 m height as a large tree and blooms from late April to mid-May (Fig. 1). Its fragrant, beautiful, large flowers and orange arils make it an excellent ornamental plant species (Fig. 1). This species has male and perfect flowers in different plants and is thus morphologically androdioecious (Liu 1984). *P. omeiensis* has been listed as Grade I in National Key Protected Wild Plants in 1999 and has been listed as Critically Endangered (CR) by the *IUCN Red List* in

D. Yu · C. Li
Sichuan Provincial Institute of Natural Resource Science, Chengdu, Sichuan, China

Emeishan Botanical Garden, Emeishan, Sichuan, China

X. Wen (✉)
Botanic Gardens Conservation International (BGCI) China Office, South China Botanical Garden, Chinese Academy of Sciences, Guangzhou, China
e-mail: wx-ying@scib.ac.cn

H. Ren (ed.), *Conservation and Reintroduction of Rare and Endangered Plants in China*, https://doi.org/10.1007/978-981-15-5301-1_11

Fig. 1 The flower, fruit, and individual of *Parakmeria omeiensis*

2007 (IUCN 2007), the *Chinese Higher Plants Red List* in 2013, and the *Red List of Magnoliaceae* in 2016 (Malin et al. 2016). The tree has also been identified as a Plant Species with Extremely Small Populations for urgent conservation in China (Ren et al. 2012).

Parakmeria omeiensis only distributes in the northeast and northwest slopes of Mt. Emei, southwest Sichuan Province, China. There are only two wild populations at the region. It grows under the canopy of the evergreen broadleaf forest (Fig. 2). Intensive forest logging is the most important threat to this species. Other threats include the difficulty in pollination, seed germination, the fragmentation of habitat, as well as the damage from monkeys. It was estimated that there were less than 100 individuals of this species in the wild.

2 Description of Reintroduction

2.1 Feasibility

Studies on the distribution and plant communities (Zhuang et al. 1993), tissue culture (Yu et al. 2011), hand-pollination, propagation technique (Yu and Li 2017; Yu et al. 2020), and floral biological characteristics of *P. omeiensis* (Yu et al. 2019) have been

Fig. 2 The habitat of *Parakmeria omeiensis*

conducted. Seeds from the wild were collected and sown at Emeishan Botanical Garden in 1989. So far, a total of 65 plants have been obtained, in which 25 plants have blossomed. About 1300 plump seeds were collected by hand-pollination (Fig. 3) and germinated at Emeishan Botanical Garden. A total of 1750 saplings have been raised during 2015–2017, including seedlings and graftings, which provided the materials for the reintroduction of *P. omeiensis* to the wild sites in Mt. Emei, southwest China.

Fig. 3 The hand-pollination of *Parakmeria omeiensis*

2.2 Implementation

Saplings were transplanted into black nursery bags (12–16 cm in diameter) filled with mixed humus and pearlites at least 1 year before they were outplanted to the wild. During transporting, these saplings were tied with ropes and wrapped with wet newspaper. With the engagement of local horses and villagers (Fig. 4), a total of 600 saplings including 440 seedlings and 160 female graftings were transplanted into the wild sites with an altitude range of 1200–1550 m, where it is similar to its original habitat. Before planting, black nursery bags for all saplings were removed. The interfaces of grafting saplings were buried into the soil to prevent them from breaking. All graftings were interplanted among seedlings. After outplanting, saplings were not watered and waited for raining. The planting plot was kept natural, i.e., no fertilizing, no clearing of other plants, etc. The only treatment was the trimming of the dense branches around the transplanted saplings for exposure to sunlight. All transplanted saplings were measured and documented before

Fig. 4 Bringing saplings to the reintroduction site

outplanting and labeled for post-planting monitoring in the next few years (Yu et al. 2020).

2.3 Post-planting Monitoring

After transplanting, all labeled plants were monitored including their survival, height, the number of new branches, crown, and soil physicochemical properties from 2016 to 2019 (Figs. 5 and 6). The dead ones were replaced and the causes of death were analyzed. The monitoring was carried out quarterly in the first year and yearly thereafter. With the well aftercare, the survival rate reached to 96.3% at the altitude of 1200–1350 m, where 480 saplings were planted, and 77.5% at the altitude of 1500–1550 m, where 120 saplings were planted. So far, there are four wild populations including two original populations and two reintroduced populations. The major reason leading to the lower survival rate (77.5%) was the wet soil with the water content of 98.8%. The wet soil facilitated the moss to grow at the base of some saplings. The moss prevented sufficient soil aeration and caused root death.

Twenty-one saplings planted at the altitude of 1500–1550 m died due to the wet and mossy soil condition. Tourists were another factor leading to plant death. Specifically, six saplings planted by the roadside were destroyed by tourists who broke the plants' branches. Our field observations indicated that transplanted

Fig. 5 Monitoring the growth of *Parakmeria omeiensis* reintroduced individuals

saplings of *P. omeiensis* grew very well in well-drained gravel soil with rich organic matters.

3 Problems and Recommendations

- Because of the harsh habitat, there might have more individuals not found in the wild. More field surveys in different seasons especially flowering season should be conducted, and high technology such as drones should be used.

Fig. 6 Reintroduced individuals are growing very well

- Mt. Emei is a famous scenic spot that attracts millions of tourists. More public education activities should be carried out to raise their awareness, accompanied by outreach materials.
- Both the original distribution areas and reintroduced sites are not far from the rhesus monkey zone with many monkeys. Patrols need to be further strengthened although the Mt. Emei Scenic Area Management Committee has paid more attention since the starting of this project.
- Both the original distribution areas and reintroduced sites are quite remote and difficult to reach, which increases the difficulties of managing and monitoring plants. All stakeholders including local villagers, local forestry institutes, and the local government should work together.

- The best method for conservation of rare and endangered plant is integrated conservation, integrating ex situ conservation (establishing an ex situ living collection that contains the entire wild genetic diversity) and in situ preservation (patrolling, etc.), combining with reintroduction to increase the wild population numbers and sizes.
- Wild plant conservation is a long-term and complicated project, which needs the cooperation among multiple disciplines and multi-sectors.

References

IUCN, Threats Authority File v. 2.1 (2007). http://www.iucnredlist.org/info/major-threats

Y.H. Liu, A preliminary study on the taxonomic system of Magnoliaceae. Acta Phytotax. Sin. **22**, 89–109 (1984)

R. Malin, B. Emily, M. Lydia, O. Sara, *The Red List of Magnoliaceae*. Revised and extended (Botanic Gardens Conservation International, Surrey, UK, 2016)

H. Ren, Q.M. Zhang, H.F. Lu, Wild plant species with extremely small populations require conservation and reintroduction in China. Ambio **41**, 913–917 (2012)

D.P. Yu, C.H. Li, Rescuing *Parakmeria omeiensis*, a critically endangered species endemic to Sichuan, China. Green Tianfu **9**, 26–27 (2017)

D.P. Yu, C.H. Li, J. Ding, et al., Preliminary report on character of seed and seed culture in vitro of *Parakmeria omeiensis*. Resour. Dev. Market **27**, 197–198 (2011)

D.P. Yu, C.H. Li, X.Y. Wen, et al., Flowering biology and breeding system of *Parakmeria omeiensis*. Guihaia **39**, 600–607 (2019)

D.P. Yu, X.Y. Wen, C.H. Li, et al., Integrated conservation for *Parakmeria omeiensis* (Magnoliaceae), a critically endangered plant species endemic to south-west China. Oryx **54**, 460–465 (2020)

P. Zhuang, R.Y. Liu, K.H. Liang, A preliminary study on phytocoenological characters of *Parakmeria omeiensis* community. Guihaia **13**, 61–69 (1993)

Reintroduction of *Hopea hainanensis* Merr. and Chun, a Wild Plant with Extremely Small Populations, in Southern China

Runguo Zang, Yi Ding, Jihong Huang, and Yue Xu

Abstract *Hopea hainanensis* Merr. and Chun (Dipterocarpaceae) mainly distributes in the tropical lowland rain forests in Hainan Island and northern Vietnam. Population size of this species decreased greatly in recent years because of extremely high value in the timber market. Thus *H. hainanensis* is listed as an endangered species in the *IUCN Red List of Threatened Species 2018* and the list *Wild Plants with Extremely Small Populations in China* in 2012. The germination rate of seeds is high after removing the seed wings. Thus, the materials for reintroduction are relatively abundant. We conducted reintroduction experiment in 2017 from August to September. The 3-year-old seedlings of *H. hainanensis* were planted in the understory of the 60-year-old secondary forest after thinning treatment or without any silviculture management. A total of 9332 individuals of *H. hainanensis* were planted. After transplantation, all individuals were tagged and measured in December 2017. Then all seedlings were re-censused in December 2018 and December 2019. Results showed that 91% of the seedlings were still alive after 2 years of transplantation. Moreover, the survival rate of reintroduced seedlings was higher after thinning treatment (93.6%) than that of control (88.5%). The long-term monitoring is still needed for assessing the effect of thinning treatment on reintroduction in the understory of the secondary forest.

Keywords *Hopea hainanensis* · Tropical lowland rain forests · Thinning treatment · Restoration · Reintroduction

---

R. Zang (✉) · Y. Ding · J. Huang · Y. Xu
Research Institute of Forest Ecology, Environment, and Protection, Chinese Academy of Forestry, Beijing, China
e-mail: zangrung@caf.ac.cn

H. Ren (ed.), *Conservation and Reintroduction of Rare and Endangered Plants in China*, https://doi.org/10.1007/978-981-15-5301-1_12

1 Introduction

Hopea hainanensis Merr. and Chun is a tall tree species belonging to the family Dipterocarpaceae (Figs. 1 and 2). It is an important timber resource for making boats and building bridges and houses due to its durable wood. It is listed as one of the five most valuable timber species in Hainan Province, Southern China. Population size of the species decreased greatly in recent years because of extremely high value in the timber market. It has been reported as an endangered species with scattered occurrence and was listed as a priority protected key wild plant of China in 1999 and the list *Wild Plants with Extremely Small Populations in China* in 2012. *The IUCN Red List of Threatened Species 2018* also categorized it as being endangered.

Hopea hainanensis mainly distributes in the tropical lowland rain forests in Hainan Island and northern Vietnam (Fig. 3). Although there is natural regeneration (Fig. 4), the population size of *H. hainanensis* decreased greatly in recent years because of illegal logging. In a typical and well-conserved forest region (108°53′–109°20′E, 18°52′–19°12′N, 500 km^2 in size) in Bawangling forest region in Hainan,

Fig. 1 The adult individuals of *Hopea hainanensis*

Fig. 2 The fruits of *Hopea hainanensis*

Fig. 3 The habitat of *Hopea hainanensis*

only 155 adult individuals (diameter at breast height > 10 cm) were found in the field investigation in 2013.

Fig. 4 The naturally regenerated seedlings of *Hopea hainanensis*

2 Description of Reintroduction

2.1 Feasibility

We conducted two field investigations in Bawangling forest region on Hainan Island in 2012 and 2017, respectively. Detailed information of distribution, habitat feature (topography, soil nutrients, canopy openness), stem size, and seedling regeneration of 60 adult trees of *H. hainanensis* was collected (Lu et al. 2020). The germination rate of seeds is high after removing the seed wings. Thus, the materials for reintroduction are relatively abundant. We selected the 60-year-old secondary forest as main reintroduction site, including *H. hainanensis*. The late-successional species are the most important component in driving ecological functioning of natural tropical forests (Zang et al. 2010). The purpose of reintroduction is to recover tropical lowland rain forests by increasing the abundance of canopy species (Swinfield et al. 2016).

2.2 Implementation

We conducted reintroduction experiment in 2017 from August to September, which is the rain season in our site (Fig. 5). We planted 3-year-old seedlings (Fig. 6) of *H. hainanensis* in the understory of the 60-year-old secondary forest after thinning treatment (A section of plots in Fig. 5) or without any silviculture management

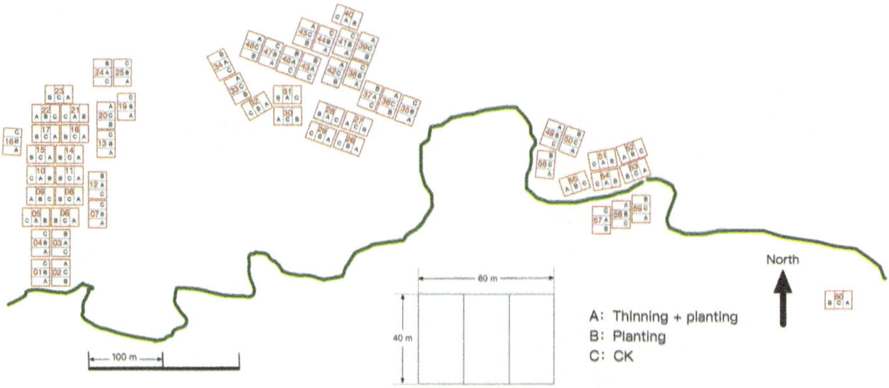

Fig. 5 The map of field experiment for reintroduction of *Hopea hainanensis* in 2017

Fig. 6 The seedlings of *Hopea hainanensis* for understory planting in Bawangling forest region, Hainan Island, South China

(B section in Fig. 5). A total of 9332 individuals of *H. hainanensis* were planted in section A and B (Fig. 5). The distance between each individual is 3 × 3 m. After planting, we applied 0.1 kg compound fertilizer to each seedling, and anti-ant medicine was also sprayed to prevent root damage from termites. We replaced the dead individuals during the first 3 months.

2.3 Post-planting Monitoring

After transplantation, all individuals were tagged and measured in December 2017. Then all seedlings were re-censused in December 2018 and December 2019 (Fig. 7). Results showed that 91% of the seedlings were still alive after 1 year of transplantation (Fig. 8). Moreover, the survival rate of reintroduced seedlings was higher after thinning treatment (93.6%) than that of control (88.5%). However, long-term monitoring is still needed for assessing the effect of thinning treatment on reintroduction in the understory of the secondary forest.

Fig. 7 Monitoring the growth of reintroduced individuals of *Hopea hainanensis* in 2018

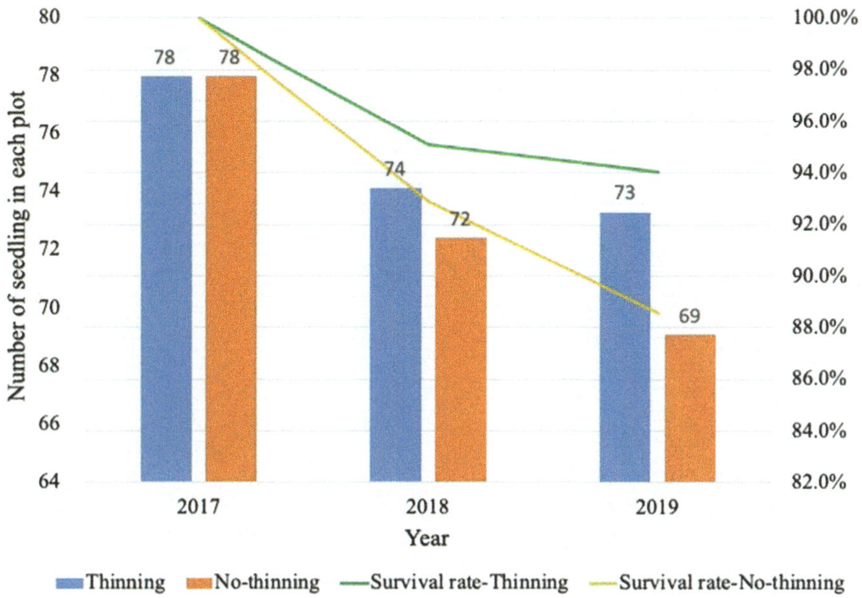

Fig. 8 The number and survival rates of reintroduced seedlings of *Hopea hainanensis* in 2018 and 2019 after different tending treatments in 2017

3 Problems and Recommendations

- The extremely high value of *H. hainanensis* on timber market makes adult trees vulnerable to illegal logging. Thus, more rigorous in situ conservation is urgently needed.
- Regeneration rate of this species was closely related to its habitat; thus the protection of habitat can facilitate their regeneration.
- The seedlings of the species might be successfully reintroduced under relatively old secondary forests or selectively logged tropical lowland rain forest if silviculture management is conducted.
- It is very important to prevent root damage from termites after transplantation.

References

X. Lu, R. Zang, Y. Ding, et al., Habitat characteristics and its effects on seedling abundance of *Hopea hainanensis*, an endangered plant with small populations. Biodivers. Sci. **28**, 289–295 (2020)

T. Swinfield, R. Afriandi, F. Antoni, et al., Accelerating tropical forest restoration through the selective removal of pioneer species. For. Ecol. Manage. **381**, 209–216 (2016)

R.G. Zang, Y. Ding, Z.D. Zhang, et al., *Ecological Base of Conservation and Restoration for the Major Functional Groups in Tropical Natural Forests on Hainan Island* (Science Press, Beijing, 2010)

Reintroduction of *Trigonobalanus doichangensis*, a Threatened Plant Species with Extremely Small Populations

Weibang Sun, Fengrong Li, and Yuan Zhou

Abstract Based on the comprehensive studies on the conservation biology of *Trigonobalanus doichangensis*, a globally threatened plant species with extremely small populations (PSESP), some 247 2-year-old saplings propagated from seeds were planted into its originally distributed site (99°44′57.5″E, 22°28′43.0″N and altitude of 1352 m) in south Yunnan province of China in September 2007. In cooperation with the local Forestry Bureau of Lancang County and a local family near the site, the necessary aftercares, monitoring, and data collecting have all been continuously carried out, around 50% of the survived plants bloomed and fruited in 2015. Twelve years after planting in September 2019, a survival rate of 46.56%, an average height increase of 7.89 m, and an average base diameter increase of 5.40 cm of *T. doichangensis* were all recorded and documented. Certainly, this successful reintroduction practice can be a template for accelerating the multiple-scale population restoration of *T. doichangensis* in China.

Keywords *Trigonobalanus doichangensis* · A threatened tree with extremely small populations · Reintroduction · Flowered and fruited · A template for population restoration

1 Introduction

The broadly circumscribed genus *Trigonobalanus* includes three species: *Trigonobalanus verticillata* from Sulawesi, Borneo, and the Malay Peninsula; *T. excelsa* from the tropical forests of Colombia, South America; and *T. doichangensis* distributed in Southern China and northern Thailand (Hsu et al.

W. Sun (✉) · F. Li · Y. Zhou
Yunnan Key Laboratory for Integrative Conservation of Plant Species with Extremely Small Populations, Kunming Institute of Botany, Chinese Academy of Sciences, Kunming, Yunnan, China
e-mail: wbsun@mail.kib.ac.cn

H. Ren (ed.), *Conservation and Reintroduction of Rare and Endangered Plants in China*, https://doi.org/10.1007/978-981-15-5301-1_13

1981). Alternatively, on the basis of the unique pollen (peroblate shape and presence of endoaperature) of *T. doichangensis*, whorled phyllotaxy with interpetiolar stipules in *T. verticillata*, and lack of bud scales in *T. excelsa*, Nixon and Crepet (1989) proposed dividing the genus into three monotypic genera: *Formanodendron*, *Trigonobalanus*, and *Colombobalanus*, respectively (Sun et al. 2011). *T. doichangensis* was firstly recorded in China in 1981 (Hsu et al. 1981), and it was listed as "a second-class protected key wild Plants of China (First Group)" (Yu 1999), evaluated as a globally endangered species (Sun et al. 2006, 2011), and as one of the 231 targeted species in a China's National Key Program: Survey and Germplasm Conservation of Plant Species with Extremely Small Populations in Southwest China (grant number: 2017FY100100) (Yang and Sun 2017; Sun et al. 2019). *T. doichangensis* is an evergreen tree narrowly distributed in the evergreen broadleaf forests in southern Yunnan and Chiang Rai in northern Thailand at an altitude of 1000–1900 m (Sun et al. 2011). In China, only the populations located in the national natural reserve in Cangyuan county of Yunnan have been well protected, while other populations/individuals have been seriously threatened due to the agricultural land expansion and cuttings for fuelwood and agricultural tools and invaded by the invasive species such as *Chromolaena odorata* and *Tithonia diversifolia* (Fig. 1, Sun et al. 2006, 2011).

2 Description of Reintroduction

2.1 *Feasibility*

We have studied the seed and seed germination biology (Zhou et al. 2003; Zheng and Sun 2008; Zheng et al. 2009; Chen et al. 2011), population biology, ecology and reproductive biology (Sun et al. 2004, 2006), cytology (Han and Sun 2005; Chen et al. 2007; Chen and Sun 2010), and genetic diversity (Sun et al. 2007) of this endangered species. We have also comprehensively summarized the progress and perspectives of conservation biology on *T. doichangensis* (Sun et al. 2011). Meanwhile, the propagation techniques on seed sowing, cuttings, and tissue culture were obtained, and more than 2000 saplings of *T. doichangensis* from four populations (Menglian, Lancang, Ximeng, and Cangyuan) from south Yunnan and one population from Chiang Rai in northern Thailand have been successfully propagated and ex situ conserved at Kunming Botanical Garden (KBG) under Kunming Institute of Botany, Chinese Academy of Sciences. Some ex situ conserved Plants at KBG have started to bear flowers and fruits since 2010.

2.2 *Implementation*

In September 2007, in cooperation with Botanic Gardens Conservation International (BGCI) and Forestry Bureau of Lancang County in China's Yunnan province,

Fig. 1 Habitat destruction of *Trigonobalanus doichangensis* caused by agricultural land expansion (**a**), cuttings for fuelwood and agricultural tools (**b**), and invasive species of *Chromolaena odorata* (**c**)

247 2-year-old individuals of *T. doichangensis* from different populations have been planted in the farmland belonging to the Banli village, Donghui town of Lancang County (Fig. 2). We were told that the farmland (reintroduced site) was used to be virgin forest, in which *T. doichangensis* was distributed. The geographical location of the reintroduction site is 99°44′57.5″E, 22°28′43.0″N, and its altitude is 1352 m above sea level.

Fig. 2 The reintroduced 2-year saplings of *Trigonobalanus doichangensis* (**a**), planting (**b**), and group photo (**c**)

Fig. 2 (continued)

2.3 *Post-planting Monitoring*

Under the supervision of Forestry Bureau of Lancang County, one family in Banli village was responsible to maintain the reintroduced plants without damage from people and animals and necessary watering and weeding. We have comprehensively monitored the survival, height, diameters of stem base, and the trunk 1.3 m above the ground (if the plants are big enough) every year from 2007 to 2011 (Fig. 3) and every 2 years from 2011 to 2019. The reintroduced plants were growing well, with about 50% of the surviving individuals having bloomed and fruited in 2015 (Fig. 4). By September 2019, the survival rate was 46.56%, the average plant height was 8.84 m (an average height increase of 7.89 m), and the average base diameter was 6.07 cm (the average increase of 5.40 cm). In summary, this reintroduction project provides a template for the multiple-scale population restoration practices for *T. doichangensis* and even other threatened woody plants in China.

3 Problems and Recommendations

- To make this reintroduction truly successful, the long-term maintenance, monitoring, and data documentation and analysis on blooming/fruiting and population regeneration and expansion of *T. doichangensis* are essential.

a

b

Fig. 3 Monitoring and data collecting of *Trigonobalanus doichangensis* in 2009 (**a**) and 2011 (**b**)

- It is also very important to get funding or project supports to continue the cooperation with local forestry departments and communities. This might be another challenge.
- More reintroductions or reinforcements for *T. doichangensis* based on the practice of this demonstration project might be urgently needed for effective conservation of the species and its populations.

Fig. 4 The flowered and fruited plant (**a**), inflorescence (**b**), and infructescence (**c**) of *Trigonobalanus doichangensis*

- To alternatively protect *T. doichangensis* (even the reintroduced plants), it might be also necessary to create the artificial plantings to meet local residents' needs for fuelwood and agricultural tools.

References

G. Chen, W.B. Sun, Ploidy variation in *Trigonobalanus verticillata* (Fagaceae). Plant Syst. Evol. **284**, 123–127 (2010)

G. Chen, W.B. Sun, C.Y. Han, et al., Karyomorphology of the endangered *Trigonobalanus doichangensis* (a.Camus) Forman (Fagaceae) and its taxonomic and biogeographic implications. Bot. J. Linn. Soc. **154**, 321–330 (2007)

H.Y. Chen, W.B. Sun, W.Q. Li, Seed storage behavior of *Formanodendron doichangensis* (Fagaceae), a rare and endangered plant. Plant Divers. Resour. **33**(5), 540–546 (2011)

C.Y. Han, W.B. Sun, Karyotype of the 4 populations of *Trigonobalanus doichangensis* (Fagaceae), a rare and endangered Plant in China. Acta Bot. Yun. **1**, 95–100 (2005)

Y.C. Hsu, C.J. Wang, C.Y. Wu, et al., *Trigonobalanus* Forman—A new recorded genus of Fagaceae in China. Acta Bot. Yun. **3**(2), 213–215 (1981)

K.C. Nixon, W.L. Crepet, *Trigonobalanus* (Fagaceae): Taxonomic status and phylogenetic relationships. Am. J. Bot. **76**(6), 828–841 (1989)

W.B. Sun, C.Y. Han, L.M. Gao, et al., Genetic diversity of the rare Asian plant, *Trigonobalanus doichangensis* (Fagaceae). Aust. J Bot. **55**, 10–17 (2007)

W.B. Sun, J. Yang, Z.L. Dao, *Study and Conservation of Plant Species with Extremely Small Populations (PSESP) in Yunnan Province, China* (Science Press, Beijing, 2019)

W.B. Sun, Y. Zhou, C.Y. Han, et al., Status and conservation of *Trigonobalanus doichangensis* (Fagaceae). Biodivers. Conserv. **15**, 1303–1318 (2006)

W.B. Sun, Y. Zhou, C.Y. Han, et al., Trigonobalanus, in *Wild Crop Relatives: Genomic and Breeding Resources, Forest Trees*, ed. by C. Kole, (Springer-Verlag, Berlin/Heidelberg, 2011), pp. 145–160

W.B. Sun, Y. Zhou, J.C. Zhao, et al., Current distribution, population attributes and biological characters of *Trigonobalanus doichangensis* in relation to its conservation. Acta Ecol. Sin. **24** (2), 352–358 (2004)

J. Yang, W.B. Sun, A new programme for conservation of plant species with extremely small populations in Southwest China. Oryx **51**(3), 396–397 (2017)

Y.F. Yu, List of national key protected wild plants (first group). Plant J. **5**, 4–11 (1999)

Y.L. Zheng, W.B. Sun, Variation in morphophysiological characters of fruits of *Trigonobalanus doichangensis* (Fagaceae) according to individual trees, populations and years. Euphytica **164**, 231–238 (2008)

Y.L. Zheng, W.B. Sun, Y. Zhou, et al., Variation in seed and seedling traits among natural populations of *Trigonobalanus doichangensis* (A. Camus) Forman (Fagaceae), a rare and endangered plant in Southwest China. New For. **37**, 285–294 (2009)

Y. Zhou, W.B. Sun, C.R. Li, Preliminary study on germination of *Trigonobalanus doichangensis*. J Wuhan Bot. Res. **1**, 73–76 (2003)

Reintroduction of *Craigia yunnanensis* to Private Plots: An Innovative Case with Help from Local Volunteers

Jing Yang, Yaling Chen, and Weibang Sun

Abstract The habitats of *Craigia yunnanensis* (Tiliaceae), an endangered tree with extremely small populations, have been mostly occupied by human settlements and farmlands, and thus population restoration in these areas might not be a practical option. Therefore, we have attempted an innovative reintroduction practice since 2017. Local volunteers, who were also landowners, adopted and planted seedlings dispersively to their private idle plots around the countryside, which helped to avoid the interference of seedling growth with their crops or houses and also increased the guarantees to the reintroduction success.

Keywords *Craigia yunnanensis* · Reintroduction · Volunteers · An innovative case · Private idle plots

1 Introduction

Craigia yunnanensis W. W. Smith and W. E. Evans is the only survived species in the genus *Craigia* (Tiliaceae), which was once widespread during the Tertiary period (Kvaček et al. 2002; Jin et al. 2009). *C. yunnanensis* is a deciduous canopy tree occupying limestone mountainous forest (Figs. 1 and 2). As the result of deforestation, most of the habitat of this species was destroyed, and the remaining habitat was occupied by human settlements. Based on our extensive survey of the natural habitats, about 20 populations and fewer than 300 individuals of this species were exited in wild. This species was listed as a second-class protected key wild plant of

J. Yang (✉) · W. Sun
Chinese Academy of Sciences (CAS), Kunming Institute of Botany, Kunming, Yunnan, China
e-mail: yangjing@mail.kib.ac.cn; wbsun@mail.kib.ac.cn

Y. Chen
Chinese Academy of Sciences (CAS), Kunming Institute of Botany, Kunming, Yunnan, China

Chinese Academy of Sciences, University of Chinese Academy of Sciences, Beijing, China

H. Ren (ed.), *Conservation and Reintroduction of Rare and Endangered Plants in China*, https://doi.org/10.1007/978-981-15-5301-1_14

Fig. 1 Big tree of *Craigia yunnanensis* and its disturbed habitat

Fig. 2 The flowers of *Craigia yunnanensis*

China in 1999 and included in China's national list of plant species with extremely small populations (PSESP) for urgent rescue (Ma et al. 2013; Sun 2013; Sun et al. 2019). *C. yunnanensis* was also included in China's National Key Program: Survey and Germplasm Conservation of PSESP in Southwest China (grant number: 2017FY100100) for the comprehensive investigations and ex situ protection (Yang and Sun 2017).

2 Description of Reintroduction

2.1 *Feasibility*

The distribution, population structure, reproduction, germination, and genetic diversity of *C. yunnanensis* were studied since 2010 (Gao et al. 2010, 2012; Yang et al. 2016; Sun et al. 2019). An extremely small population with only seven individuals of this species was found in Jiangdong town, Dehong county of Yunnan province, China (Sun et al. 2019). This population is a typical example to demonstrate the disturbance of human settlement to this species. Like most other populations, several villages have occupied the whole habitat, and survived individuals were distributed around cottages or in the private plots. Since the habitat was mostly replaced by private farmlands and economy forests, reintroduction in planned and settled plots was not an option. Therefore, we planned an innovative reintroduction and try to find if it was possible to recover the population by reintroducing the seedlings dispersively to private idle plots around the countryside.

2.2 *Implementation*

We collected seeds of this population and with the help of a local villager. About 150 seedlings of *C. yunnanensis* were obtained in 2016. In 2017, the forestry station of Jiangdong town introduced our plan to local villagers. The threatened status and importance of conservation of *C. yunnanensis* were discussed and volunteers were recruited. We informed the requirements and notes as follows: (1) we needed the volunteers to plant the seedlings into their private idle plots by themselves; (2) the volunteers should notice that the future growth of the seedlings should not disturb the growth of other crops in their private plots; and (3) we would not criticize, judge, or ask for any compensation if the seedlings were dead or had been damaged by the volunteers. We emphasized that our intention was to recruit the villager's private idle plots as the reintroduction plots, of which only the volunteers know where the locations and trees can be planted freely. In the end, nine volunteers, who are mostly forestry rangers, and the forestry station of Jiangdong town received and planted the seedlings.

2.3 *Post-release Monitoring*

In 2018, we surveyed and recorded the GPS coordinates of live seedlings. Altogether, 74 seedlings were alive. One volunteer planted the seedlings beside the village's roads with the intention of growing shade trees, but they were lost to the road reconstructions. One volunteer fenced the seedlings as a protection. Some

Fig. 3 The volunteers and the reintroduced seedlings in their private plots

volunteers planted the seedlings at the edges of their farmlands, because it is convenient to look after the seedlings when they take care of the crops (Fig. 3). Of the resurvey in 2019, a total of 50 individuals were still alive.

3 Problems and Recommendations

- The commitment to caring for reintroduced seedling is totally dependent on the passion of the volunteers.
- The survival rate of seedlings cannot be guaranteed.
- The cooperation of local forestry departments was strongly recommended.
- The private idle plots were less disturbed habitat.
- The reintroduced individuals could enlarge the distribution area.

References

Z.R. Gao, C.Q. Zhang, Z.Q. Han, et al., Pollination biology and breeding system of *Craigia yunnanensis* in fragmented habitat. Chin. J. Ecol. **31**, 2217–2224 (2012)

Z.R. Gao, C.Q. Zhang, R.I. Milne, Size-class structure and variation in seed and seedling traits in relation to population size of an endangered species *Craigia yunnanensis* (Tiliaceae). Aust. J. Bot. **58**, 214–223 (2010)

J. Jin, T.M. Kodrul, W. Liao, et al., A new species of *Craigia* from the Eocene Changchang formation of Hainan Island, China. Rev. Palaeobot. Palynol. **155**, 80–82 (2009)

Z. Kvaček, S.R. Manchester, R. Zetter, et al., Fruits and seeds of *Craigia bronnii* (Malvaceae—Tilioideae) and associated flower buds from the late Miocene Inden formation, lower Rhine Basin, Germany. Rev. Palaeobot. Palynol. **119**, 311–324 (2002)

Y.P. Ma, G. Chen, R.E. Grumbine, et al., Conserving plant species with extremely small populations (PSESP) in China. Biodivers. Conserv. **22**, 803–809 (2013)

W.B. Sun, *Conserving Plant Species with Extremely Small Populations (PSESP) in Yunnan: A Practice and Exploration* (Yunnan Science and Technology Press, Kunming, 2013)

W.B. Sun, J. Yang, Z.L. Dao, *Study and Conservation of Plant Species with Extremely Small Populations (PSESP) in Yunnan Province, China* (Science Press, Beijing, 2019)

J. Yang, Z.R. Gao, W.B. Sun, et al., High regional genetic differentiation of an endangered relict plant *Craigia yunnanensis* and implications for its conservation. Plant Divers. **38**, 221–226 (2016)

J. Yang, W.B. Sun, A new programme for conservation of plant species with extremely small populations in south-West China. Oryx **51**, 396–397 (2017)

Reintroduction of *Myricaria laxiflora* (Franch.) P.Y. Zhang et Y.J. Zhang, a Critically Endangered Shrub, in Central China

Zhiqiang Xiao and Mingxi Jiang

Abstract *Myricaria laxiflora* (Tamaricaceae) is an endangered and endemic riparian shrub along Yangtze River in Three Gorges Reservoir Area (TGRA) of Central China. Because of the construction of the Three Gorges Dam (TGD) project, most of natural distribution sites of *M. laxiflora* were destroyed. We have established six reintroduction sites for this species in Chongqing Municipality and Hubei Province. The growth rhythm of transplanted populations was basically consistent with its origin population. The reintroduction and ex situ populations of *M. laxiflora* can blossom and bear fruit normally, but not produce offspring, so there are renewal barriers. It is necessary to continue monitoring and managing the reintroduction populations.

Keywords Three Gorges Reservoir Area · Riparian zone · *Myricaria laxiflora* · Ex situ · Reintroduction

1 Introduction

Myricaria laxiflora (Franch.) P.Y. Zhang et Y.J. Zhang (Tamaricaceae) is an endangered and endemic riparian shrub (Fig. 1). It was considered to be restricted along Yangtze River in Three Gorges Reservoir Area (TGRA) of Central China. *M. laxiflora* prefers the alluvial soil with sand and gravel in subtropic climate. The soil is weak alkalinity and has low content of nutrient but moderate salt. Due to the construction of the Three Gorges Dam (TGD) project, all distribution sites of *M. laxiflora* were flooded in TGRA. However, in December 2017, three wild populations (Zhijiang, Yidou, and Yichang) of this species were found at western Hubei Province along the Yangtze River valley downstream of the TGD (Fig. 2). It was estimated that there were about 20,000 individuals in the wild.

Z. Xiao · M. Jiang (✉)
Chinese Academy of Sciences, Wuhan Botanical Garden, Wuhan, China
e-mail: mxjiang@wbgcas.cn

H. Ren (ed.), *Conservation and Reintroduction of Rare and Endangered Plants in China*, https://doi.org/10.1007/978-981-15-5301-1_15

Fig. 1 The individuals of *Myricaria laxiflora*

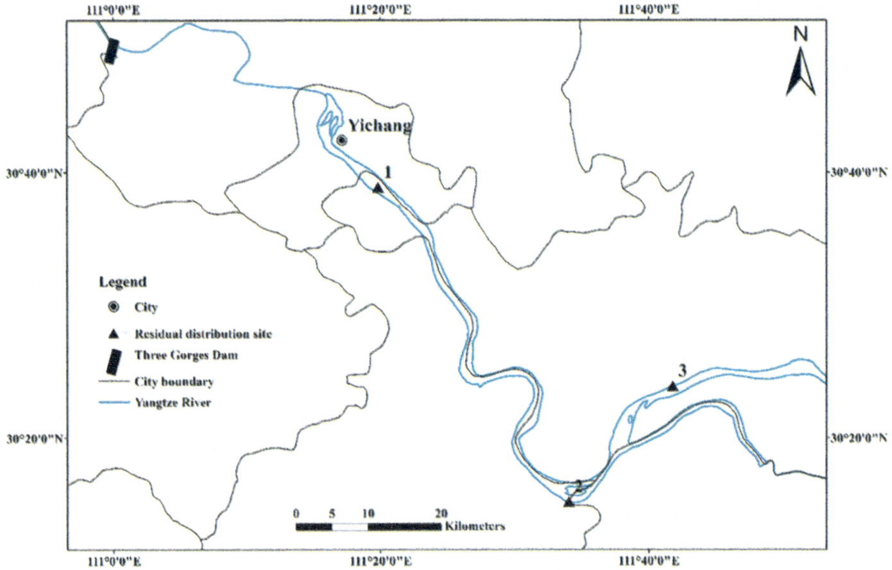

Fig. 2 Schematic diagram of natural populations of *Myricaria laxiflora*. 1, Yanzhiba in Yichang; 2, Guanzhou in Yichang; 3, Dongshi Town in Zhijiang

2 Description of Reintroduction

2.1 *Feasibility*

Researchers at Wuhan Botanical Garden, Chinese Academy of Sciences (WBGCAS) have studied the distribution, biological characteristics, community ecology, pollination biology, genetic diversity, propagation, and conservation biology of this species since the 1990s (Wu et al. 1998; Shen et al. 1999; Wang et al. 2003; Li et al. 2003; Tao et al. 2004; Bao et al. 2010; Xu et al. 1999; Chen et al. 2005a, b; Chen and Xie 2008; Han et al. 2008). According to the principle of similar habitats, convenient management and multisite preservation, we established six reintroduction sites for this species in Chongqing Municipality and Hubei Province during 2003–2004 (Fig. 3). Seed germination and cutting propagation tests were successful during 1996–1997 (Xu et al. 1999).

2.2 *Implementation*

Myricaria laxiflora below the submerged line were transplanted to reintroduction sites during 2003–2004. The sizes of all reintroduction populations were greater than 1000 individuals (Chen et al. 2005a). New habitats were selected at higher ground, and large boulders were used to press and surround the seedlings to reduce the impact of major floods on their initial settlement. After flooding, seedlings should be straightened timely. The transplants were watered shortly after transplanted. We also

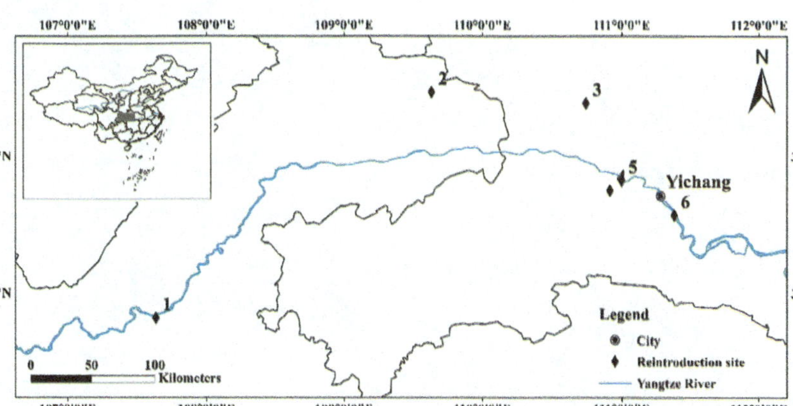

Fig. 3 Schematic diagram of reintroduction populations of *Myricaria laxiflora.* 1, Baishui reservoir in Fengdou County; 2, Daxin river in Wuxi County; 3, Xiangxi river in Xinshan County; 4, Sixi scenic spot in Zigui County; 5, Dam area of Three Gorges Project; 6, Yichang Yangtze River Highway Bridge

successfully established an ex situ site at Wuhan Botanical Garden, Chinese Academy of Sciences during 2003–2004.

2.3 Post-planting Monitoring

After transplantation, we monitored the survival, number of plant branches, flowering rate, seed setting rate, number of seedlings, plant height, and basal diameter (Fig. 4). We compared the temperature, light, soil, and other ecological conditions between ex situ and original populations. At present, the reintroduced population has been preliminarily successful (Chen et al. 2005a). Based on the growth from two transplanted populations of Xingshan and Wuxi, there were no significant differences in survival rates. The survival rate of ex situ population was over 80% in the same year. Some individuals blossomed and bore fruit. The growth rhythm was basically consistent with its origin population. Experimental studies found that both field reintroduction and ex situ conservation were faced with technical bottlenecks. Moreover, the reintroduction and ex situ populations of *M. laxiflora* can blossom and bear fruit normally, but not produce offspring, so there are renewal barriers.

Fig. 4 The reintroduction site

3 Problems and Recommendations

- Some remaining populations may be present at the area upstream of the Three Gorges Dam, so more survey should be done.
- Long-term continuous monitoring and management scientifically will contribute to the success of the reintroduction project.
- A proper management of the remaining populations and the existing ex situ populations can minimize the loss of genetic resources of *M. laxiflora*.

References

D.C. Bao, Z.J. Lu, M.X. Jiang, et al., Population structure and dynamics of remanent *Myricaria laxiflora* downstream from the Three Gorges Dam. J. Wuhan Bot. Res. **28**, 711–717 (2010)

F.Q. Chen, Z.Q. Xie, Physiological and biochemical characteristics of *Myricaria laxiflora*, an endangered species in the Three Gorges Reservoir Area. Guihaia **28**, 367–372 (2008)

F.Q. Chen, Z.Q. Xie, G.M. Xiong, et al., Reintroduction and population reconstruction of an endangered plant *Myricaria laxiflora* in the Three Gorges Reservoir area, China. Acta Ecol. Sin. **25**, 1811–1817 (2005a)

F.Q. Chen, G.M. Xiong, Z.Q. Xie, Effects of density on seedling survival and growth of an endangered species *Myricaria laxiflora*. Biodivers. Sci. **13**, 332–338 (2005b)

M.D. Han, L. Yang, S.M. Chen, X.Y. He, Cuttage of an endangered medicinal plant *Myricaria laxiflora* in the three-gorges reservoir region. J. Southwest China Normal Univ.. (Natural Science Edition) **33**, 82–85 (2008)

Z.Z. Li, C.H. Wang, T.Q. Xu, et al., Conservation genetics of the endemic species *Myricaria laxiflora* (Tamaricaceae) in the three-gorges reservoir area, Hubei. Biodivers. Sci. **11**, 109–117 (2003)

Z.H. Shen, Y.X. Jin, J.Q. Wu, et al., An contrast on the characteristics of soil in the natural habitat and re-site reserve of two endemic species of the Three Gorges reservoir region of Yangzi River. J. Wuhan Bot. Res. **17**, 46–52 (1999)

Y. Tao, J.Q. Li, M.X. Jiang, et al., Study on diversity of structural characters of *Myricaria laxiflora*. J. Wuhan Bot. Res. **22**, 315–322 (2004)

Y. Wang, J.Q. Wu, Y. Tao, et al., Natural distribution and ex situ conservation of endemic species *Myricaria laxiflora* (Tamaricaceae) in the Three-Gorges Reservoir area, Hubei. J. Wuhan Bot. Res. **11**, 109–117 (2003)

J.Q. Wu, Z.E. Zhao, Y.X. Jin, et al., Investigation and study on the endemic plant *Myricaria laxiflora* in the three-gorge reservoir area. J. Wuhan Bot. Res. **16**, 111–116 (1998)

H.Z. Xu, Y.X. Jin, Z.E. Zhao, et al., Propagation of *Myricaria laxiflora* endemic to Three-Gorges Area. Res. Environ. Yangtze Basin. **8**, 158–161 (1999)

Reintroduction of *Bretschneidera sinensis* Hemsl., an Endangered Endemic Tree

Hongfeng Chen, Qi Qiao, Hailin Liu, Xiangying Wen, and A. J. Harris

Abstract Here, we describe ongoing, successful conservation efforts for the tree species, *Bretschneidera sinensis* Hemsl., led by South China Botanical Garden. *B. sinensis* is a dioecious, deciduous tree species with a limited range in the subtropics of Vietnam, Thailand, and Southern China. It comprises the sole species within Bretschneideraceae (Brassicales). The species is a relict of the ancient boreotropical flora and exhibits a classic pattern of biodiversity distribution, in which some species-poor Asian endemic lineages had wider distributions in the Northern Hemisphere during the Cenozoic. The status of this species as a relict and its dioecious life history and extrinsic, anthropogenic factors likely all contribute to its rarity and, consequently, its rating as "endangered" by the International Union for Conservation of Nature and "category I protected" by the Chinese government. At South China Botanical Garden, we have led efforts to conserve this species through theoretical and applied research as well as public outreach and policy recommendations. In particular, we have successfully germinated over 6000 seedlings of this species representing six natural populations, and we have introduced 500 of these seedlings into Nankun Mountain Natural Reserve to reinvigorate the existing population there. Seedlings also exhibit healthy development in cultivation at South China Botanical Garden. Therefore, we recommend to continue and improve upon our integrative restoration plan for this species, through which we will lead in situ and ex situ conservation efforts alongside public outreach and education, simultaneously seeking input from stakeholders and providing local and national policy recommendations. We believe that *B. sinensis* has great horticultural potential due to its wide ecological tolerances and attractive foliage, fruits, and flowers and also

H. Chen (✉) · Q. Qiao · H. Liu · A. J. Harris
South China Botanical Garden, Chinese Academy of Sciences, Guangzhou, China
e-mail: h.f.chen@scib.ac.cn

X. Wen
Botanic Gardens Conservation International (BGCI) China Office, South China Botanical Garden, Chinese Academy of Sciences, Guangzhou, China
e-mail: wx-ying@scib.ac.cn

H. Ren (ed.), *Conservation and Reintroduction of Rare and Endangered Plants in China*, https://doi.org/10.1007/978-981-15-5301-1_16

comprises an important model for evolutionary research, due to its classic pattern of biodiversity distribution.

Keywords *Bretschneidera sinensis* · Relict · In situ · Ex situ · Conservation

1 Introduction

Bretschneidera sinensis Hemsl. is a deciduous, dioecious tree that occurs in eastern Asia and is the sole species within the Bretschneideraceae family (Fig. 1). The species has been listed as "Endangered on the IUCN Red List of Threatened Species (2013)" and as a Category-I species in the "Key list of Protection of Wild Plants in China" (Yu 1999). *B. sinensis* exhibits a common pattern of biodiversity distribution in which a large number of modern-day species-poor lineages are endemic in eastern Asia (e.g., *Dipteronia* Oliv., *Davidia* Baill., and *Eucommia* Oliv.) but once had wider distributions throughout the Northern Hemisphere during the Cenozoic as constituents of the boreotropical flora (Wolfe 1975, 1977, 1985; Manchester et al. 2009; Shang et al. 2016). Thus, *B. sinensis* is an important model system for studies for evolutionary processes of such lineages as well as for demographics of threatened endemics. *B. sinensis* also has an aesthetically pleasing shape as a landscape tree, and it bears beautiful flowers and fruits, on account of which it is highly valued as an ornamental plant. It is also valued for its straight trunk, good ground coverage, and strong capacity to grow under a wide variety of environmental conditions.

Bretschneidera sinensis occurs at elevations of 300–1700 m south of the Yangtze River in mainland China, Taiwan, and in Thailand, Laos, Myanmar, and Vietnam

Fig. 1 An individual of Bretschneidera sinensis Hemsl.

Fig. 2 The habitat of Bretschneidera sinensis Hemsl.

(Lu and Boufford 2005). It is a broadleafed forest species that primarily grows in evergreen deciduous and broadleaf mixed forest biomes with such species as *Castanopsis eyrei* (Champ. ex Benth.) Hutch., *Cyclobalanopsis glauca* (Thunb.) Oerst., *Choerospondias axillaris* (Roxb.) B.L. Burtt & A.W. Hill, *Alniphyllum fortunei* (Hemsley) Makino, *Nyssa sinensis* Oliv., *Phoebe* Nees spp., and others. Young trees of *B. sinensis* are heliophobous and hydrophilic, and mature trees are relatively heliophilous (Qiao et al. 2010). The species has an edaphic preference of red soil, especially the yellow, hydrated form, and develops deep roots that make it robust against strong winds (Lu 1999; Wu et al. 2003; Lu and Boufford 2005). We found that the existing wild populations of *B. sinensis* are quite small and contain scattered individuals and that its habitat is highly fragmented in China (Fig. 2). Female trees are especially rare and produce only a few capsular fruits annually, each bearing one to four seeds. Thus, it is very difficult for populations of the species to renew themselves naturally. The wild populations of *B. sinensis* have also been plundered for a long time, especially to make way for agricultural expansion.

Consequently, natural populations of *B. sinensis* have been greatly reduced in recent decades and are at a high risk of local extinction.

2 Description of Reintroduction

2.1 Feasibility

In order to prevent global extinction of this species, South China Botanical Garden has undertaken work on its conservation for more than a decade. Since 2007, we have studied the natural distribution of the species; recommended its conservation status; determined its ecological and biological characteristics including genetic diversity, seed germination characteristics, and its light, temperature, and water requirements; as well as developed a tissue culture strategy for the species (Qiao et al. 2011; Wang et al. 2018; Zhang et al. 2019). Through our research, we have applied multiple methods to explore the factors that have led to its present-day endangered status, and, more recently, we have successfully reintroduced this species to the wild. We have also grown some individuals within the living research collections in South China Botanical Garden. We have observed that it grows well in cultivation, so we expect that our planned ex situ conservation efforts will yield positive results. We also carried out a pollination study of wild populations of *B. sinensis* at Nankun Mountain Natural Reserve to better understand the capacity of the species to maintain or expand its populations (Qiao et al. 2012).

2.2 Implementation

We collected seeds of *B. sinensis* representing six naturally occurring populations from Guangdong, Hunan, and Jiangxi provinces of China. For seed germination, we adopted two different methods: direct seeding and prior cutting of the seed coats. The cutting of seed coats in this species may promote seed germination (Qiao et al. 2009). Our efforts at seed germination have resulted in more than 6000 seedlings cultivated at South China Botanical Gardens and Conghua experimental station. Using these seedling resources, we have cooperated with local governments in carrying out in situ conservation of *B. sinensis*. Through these efforts, we reintroduced more than 500 seedlings of the species to the Nankun Mountain Natural Reserve in 2008, in order to improve the age structure of the naturally occurring population there and promote population recovery (Fig. 3). Additionally, we are also carrying out ex situ conservation within South China Botanical Garden and Dongguan Arboretum.

Fig. 3 Documentation of growth status of reintroduced individuals

2.3 Post-planting Monitoring

After in situ and ex situ plantings, we have conducted posttransplantation monitoring. At present, both the in situ and ex situ seedlings are growing well. Our field and experimental studies have shown that there are both intrinsic and extrinsic factors that are responsible for the current endangered status of *B. sinensis*. Intrinsic factors are primarily that the species exhibits low viability and fecundity possibly resulting from its relict (or paleoendemic) nature. Extrinsic factors are especially habitat fragmentation and vandalism.

3 Problems and Recommendations

- Track and monitor natural populations and reintroduced populations with sufficient personnel for recording field observations and data entry.
- Better understand the dynamics of dioecy in evolutionary fitness and population recovery of this species.
- Strengthen public outreach and education, especially to avoid unintentional, further damage to natural and reintroduced populations.
- Make an integrative species recovery plan involving theoretical and applied research, public outreach, and policy recommendations to increase population number and size.

- Develop and lead a cooperative conservation program that engages all stakeholders, including farmers, scientists, and the local government, to ensure success for the species.

References

A.M. Lu, *The Geography of Spermatophytic Families and Genera* (Science Press, Beijing, 1999), pp. 358–364

L.L. Lu, D.E. Boufford, Bretschneideraceae, in *Flora of China (Vol. 8 Brassicaceae through Saxifragaceae)*, ed. by Z. Y. Wu, P. H. Raven, D. Y. Hong, vol. 8, (Missouri Botanical Garden Press, St. Louis, 2005), p. 197

S.R. Manchester, Z.D. Chen, A.M. Lu, et al., Eastern Asian endemic seed plant genera and their paleogeographic history throughout the northern hemisphere. J. Syst. Evol. **47**(1), 1–42 (2009)

Ministry of Environmental Protection of the People's Republic of China. The IUCN Red List in China Angiosperm. 2013

Q. Qiao, H.F. Chen, F.W. Xing, et al., Seed germination protocol for the threatened plant species *Bretschneidera sinensis* Hemsl. Seed science and technology. Seed Sci. Technol. **37**, 70–78 (2009)

Q. Qiao, X.S. Qin, F.W. Xing, et al., Death causes and conservation strategies of the annual regenerated seedlings of rare plant, *Bretschneidera sinensis*. Acta Ecol. Sin. **31**(16), 4709–4716 (2011)

Q. Qiao, F.W. Xing, H.F. Chen, et al., Characteristics of community and in situ conservation strategy of *Bretschneidera sinensis* in Nankun Mountain, Guangdong Province. Acta Bot. Boreal-Occident Sin. **30**(2), 0377–0384 (2010)

Q. Qiao, H.F. Chen, F.W. Xing, et al., Pollination ecology of *Bretschneidera sinensis* (Hemsley), a rare and endangered tree in China. Pak. J. Bot. **44**(6), 1897–1903 (2012)

K.K. Shang, K. Song, L.J. Da, Ecology and relict tertiary deciduous trees in subtropical China, in *Vegetation Structure and Function at Multiple Temporal and Conceptual Scales*, ed. by E. O. Box, (Springer, New York, 2016), pp. 149–167

M.N. Wang, L. Duan, Q. Qiao, et al., Phylogeography and conservation genetics of the rare and relict *Bretschneidera sinensis* (Akaniaceae). PLoS One **13**(1), e0189034 (2018)

J.A. Wolfe, Some aspects of plant geography of the northern hemisphere during the late cretaceous and tertiary. Ann. Mo. Bot. Gard. **6**(2), 264–279 (1975)

J.A. Wolfe, Paleocene floras from the Gulf of Alaska region. Geol. Surv. Prof. Pap. **977**, 1–108 (1977)

J.A. Wolfe, Distribution of major vegetational types during the tertiary, in *The Carbon Cycle and Atmospheric CO_2: Natural Variations Archean to Present*, ed. by E. T. Sundquist, W. S. Broecker, (American Geophysical Union, Washington, DC, 1985), pp. 357–375

Z.Y. Wu, A.M. Lu, Y.C. Tang, et al., *The Families and Genera of Angiosperms in China: A Comprehensive Analysis* (Science Press, Beijing, 2003), p. 702

Y.F. Yu, The milestone of the work of protecting wild plants of China-the list of Chinese National Protected Wild Plants (1th). Plant **5**, 3–11 (1999)

S. Zhang, H.F. Chen, J.X. Liu, et al., Effects of drought on physiological responses of *Bretschneidera sinensis*. Int. J. Agric. Biol. **18**, 226–233 (2019)

In Vitro Propagation and Reintroduction of *Paphiopedilum wardii* Sumerh.

Songjun Zeng, Kunlin Wu, Zhilin Chen, and Jun Duan

Abstract *Paphiopedilum wardii* Sumerh. is an endangered and rare terrestrial orchid threatened with extinction in the wild due to anthropogenic activities. Asymbiotic germination provides a useful way for ex situ conservation, for reintroduction in the wild, and for commercial demand. The 180-day seeds after pollination had highest germination rate with 65.33%, and 1/2MS supplemented with 0.5 mg l^{-1} NAA, 10% coconut water, and 1.0 g l^{-1} activated charcoal was a suitable medium. Hyponex N026 medium containing 1.0 mg l^{-1} NAA, 1.0 g l^{-1} peptone, 10% CW, and 1.0 g l^{-1} AC was most suitable for subculture of plantlets. Hyponex N016 medium containing 1.0 mg l^{-1} NAA, 1.0 g l^{-1} peptone, 100 g l^{-1} banana homogenate, and 1.0 g l^{-1} AC was suitable for plantlets' growth in vitro. The plantlets 5 cm in height were transplanted in mixture media [2: 1: 1 (v/v) Zhijing stone for orchids/sieved peat/shattered fir bark]. Survival rate of 92.33% was obtained after 180 days in a greenhouse. 10,000 plantlets in vitro were obtained. We used the 1000 plantlets propagated in vitro to reintroduce into one habitat and two sites outside the species range. The 180-day transplanted seedlings were reintroduced into habitat at Gaoligong Mountain in Yunnan or alien forest habitats at Yangchun and Guangzhou in Guangdong, which had a higher survival percentage than 90-day or 360-day transplanted seedlings. About 2 years after transplantation, around 30% of plants that survived flowered from seedlings transplanted for 360 days at all three reintroduction locations. At Gaoligong Mountain, Lushui, 32% of flowering plants produced fruits and seeds.

Keywords *Paphiopedilum wardii* Sumerh. · Asymbiotic seed germination · Implementation · Reintroduction · Post-planting monitoring

S. Zeng (✉) · K. Wu · J. Duan
South China Botanical Garden, Chinese Academy of Sciences, Guangzhou, China
e-mail: zengsongjun@scib.ac.cn; wkl8@scbg.ac.cn; duanj@scib.ac.cn

Z. Chen
South China Botanical Garden, Chinese Academy of Sciences, Guangzhou, China

Horticulture Research Institute, Guizhou Academy of Agricultural Sciences, Guiyang, China

H. Ren (ed.), *Conservation and Reintroduction of Rare and Endangered Plants in China*, https://doi.org/10.1007/978-981-15-5301-1_17

1 Introduction

Paphiopedilum wardii Sumerh. is an endangered and rare orchid distributed in Lushui county and northwest of Baoshan city, western Yunnan, China, and in Northern Myanmar (Cribb 1998; Liu et al. 2009). It flowers from December to March (Fig. 1). It grows in grassy and bushy places on wooded slope or forest margins along valleys at an elevation of 1200 and 2500 m (Fig. 2). It is a high-value ornamental plant evidenced by 92 combinations registered by the Royal Horticultural Society as of March 1, 2020. The wild populations of *P. wardii* are under the threat of extinction in the wild due to anthropogenic activities, and it is listed in the Convention on International Trade in Endangered Species of Wild Fauna and Flora (CITES) such that their trade is prohibited. It is now listed as Critically Endangered (CR) in the China Species Red List by the Chinese Government. However, owing to a large number of propagations in vitro and cultivation by South China Botanical Garden, which has been applied to commercial production, a substantial population exists outside wild habitats (Fig. 3).

Fig. 1 The flowering individual of *Paphiopedilum wardii* Sumerh.

Fig. 2 The habitat of *Paphiopedilum wardii* Sumerh. in Yunnan

2 Description of Reintroduction

2.1 Feasibility

We have studied the distribution, conservation status, ecological and biological characteristics, reproductive biology, tissue culture (Zeng et al. 2012), and horticulture of *P. wardii* since 2003 (Zeng et al. 2012). An effective system of propagation in vitro was established from asymbiotic seed germination. The 180-day seeds after pollination had highest germination rate with 65.33%, and 1/2MS supplemented with 0.5 mg l^{-1} NAA, 10% coconut water, and 1.0 g l^{-1} activated charcoal was a suitable medium. Hyponex N026 medium containing 1.0 mg l^{-1} NAA, 1.0 g l^{-1} peptone, 10% CW, and 1.0 g l^{-1} AC was most suitable for subculture of plantlets. Hyponex N016 medium containing 1.0 mg l^{-1} NAA, 1.0 g l^{-1} peptone, 100 g l^{-1} banana homogenate, and 1.0 g l^{-1} AC was suitable for plantlets' growth in vitro. The plantlets 5 cm in height were transplanted in mixture media [2: 1: 1 (v/v) Zhijing

Fig. 3 The commercial production population of *Paphiopedilum wardii* Sumerh. in SCBG

stone for orchids/sieved peat/shattered fir bark]. Survival rate of 92.33%was obtained after 180 days in a greenhouse. One thousand plantlets were conducted in field establishment and habilitation at three locations: (1) Gaoligong Mountain (98° 20′ E, 25° 50' N, habitat of *P. wardii* at an elevation of 1800 m) in Lushui county, Yunnan province; (2) Ehuangzhang Nature Reserve (111°30′E, 21°55'N, at an elevation of 295 m) in Yangchun, Guangdong province (Fig. 4); and (3) Huolu Mountain (113°38′E, 23°18'N, at an elevation of 85 m) in Guangzhou, Guangdong province in 2008. The plantlets were transplanted into the three locations in April 2008. (Zeng et al. 2012).

2.2 Implementation

During the acclimation period in three reintroduction locations, the plantlets were watered on days 1, 2, 7, 14, ,and 30, but were not watered thereafter. The quadrats were not fenced, fertilized, or mulched. We observed and monitored the survival rate and growth status semiannually.

Fig. 4 Reintroduction of *Paphiopedilum wardii* Sumerh. in Ehuangzhang Nature Reserve

2.3 Post-planting Monitoring

After reintroduction, we monitored the survival and growth status of all transplants in three locations from 2008 to 2017. About 1 year after transplanting, the highest survival rate was 65% at Ehuangzhang, Yangchun, from seedlings transplanted for 180 days, which was significantly higher than at the other two transplantation locations. About 2 years after transplantation, the highest survival rate was 60.33% at Ehuangzhang, Yangchun, from seedlings transplanted for 180 days, which was also significantly higher than other treatments. Around 30% of plants that survived flowered from seedlings transplanted for 360 days at all three reintroduction locations. At Gaoligong Mountain, Lushui (Fig. 5), 32% of flowering plants produced fruits and seeds; in contrast, although there were many flowering plants transplanted in the other two habitats, no plants produced fruits or seeds maybe because of pollinators.

3 Problems and Recommendations

- There may be some remaining populations in other area on the China-Myanmar border, which need further investigate. The genetic diversity of all surviving *P. wardii* individuals should be analyzed by molecular tools to afford protection to all germplasm resources with different genetic diversities.

Fig. 5 Flowering of plant reintroduced by *Paphiopedilum wardii* Sumerh.

- A new population established outside of the original habitat. The lack of effective pollinators could be a major limiting factor for reproductive success for full recovery of a self-sustaining population. In addition, the seeds could not successfully germinate because of lack of symbiotic fungi.
- The best method for conservation of *P. wardii* is in situ preservation, which can only be resolved by legislation, public education, and suitable economic compensation. Ex situ conservation or reintroduction can be used as a helpful tool to conserve biodiversity, but it is difficult and expensive.

References

P. Cribb, *The Genus Paphiopedilum*, 2nd edn. (National History Publications, Malaysia Borneo, 1998)

Z.J. Liu, S.C. Chen, L.J. Chen, et al., *The Genus Paphiopedilum in China* (Science Press, Beijing, 2009), pp. 13–349

S.J. Zeng, K.L. Wu, J.A. Teixeira da Silva, et al., Asymbiotic seed germination, seedling development and reintroduction of *Paphiopedilum wardii* Sumerh. An endangered terrestrial orchid. Sci. Hortic. **138**, 198–209 (2012)

Reintroduction and Assisted Colonization of *Paphiopedilum spicerianum*: An Orchid with Extremely Small Populations in China

Jiangyun Gao, Xuli Fan, and Wenke Yang

Abstract *Paphioepdilum spicerianum* is a species of wild plant species with extremely small populations (PSESP) in China, and only a single population with ten mature individuals was found in Yunnan Province. To save this PSESP orchid, following the idea of orchid integrated conservation and based on our previous study on mycorrhizal fungi, pollination ecology, seed storage, and developing of asymbiotic seed germination system, we conducted reintroduction and assisted colonization of *P. spicerianum* since 2015. The results of post-release monitoring showed that the survival ratio of reintroduction was much lower than assisted colonization (40% vs. 80%) after 2 years in the end of 2017, suggesting that it is hard to establish a reintroduced population in its original habitat because it is seriously threatened by many factors and assisted colonization could be a better way to save this species. However, it still faces multiple practical problems to establishing a self-sustaining population of *P. spicerianum*, and the problems and possible solutions are discussed in this study.

Keywords Assisted colonization · Orchid conservation · *Paphiopedilum* · PSESP in China · Reintroduction

1 Introduction

Paphiopedilum spicerianum is a rare terrestrial orchid with a historic distribution in India, Bhutan, Myanmar, and South China (Rankou and Molur 2015). The population trend is decreasing, and the abundance of the species has been significantly

J. Gao (✉) · X. Fan · W. Yang
Yunnan Key Laboratory of Plant Reproductive Adaptation and Evolutionary Ecology, Yunnan University, Kunming, China

Laboratory of Ecology and Evolutionary Biology, Yunnan University, Kunming, China
e-mail: jiangyun.gao@ynu.edu.cn

H. Ren (ed.), *Conservation and Reintroduction of Rare and Endangered Plants in China*, https://doi.org/10.1007/978-981-15-5301-1_18

Fig. 1 The habitat of *Paphiopedilum spicerianum* located in a seasonal riverbank

reduced in recent decades. Although this orchid is assessed as Endangered (EN) globally on the IUCN Red List (Rankou and Molur 2015), in China only a single population with about ten mature individuals was found and confirmed in 2006 (Ye and Luo 2006). It has been listed as a wild plant species with extremely small populations (PSESP) in China, which are defined as species with low population numbers due to recent and serious human disturbance and exclude naturally rare species (State Forestry Administration of China 2012). Reintroduction is favored as a conservation strategy for PSESP plants (Ren et al. 2012). *P. spicerianum* were subterrestrial plants that grow on steep river banks (Fig. 1), and their habitat is seriously threatened due to the surrounding monoculture of coffee plantations. It flowered from the middle of Oct. to the end of Nov. (Fig. 2), and fruits

Fig. 2 Plants of *Paphioepdilum spicerianum* are flowering

matured in Nov. of next year. Based on the studies of its pollination ecology, mycorrhizal fungi diversity (Han et al. 2016), and in vitro seed germination (Chen et al. 2015), from June 2015, we conducted reintroduction in its original habitat and assisted colonization in Tianzi Reserve at Bulangshan, where it is near the border between China and Myanmar with the elevation ca. 400 m higher than the original habitat at the same time.

2 Description of Reintroduction

2.1 Feasibility

The initial aim of this project was to enhance the only population of *P. spicerianum* found in China and to establish a new population in a higher elevation site. For *P. spicerianum*, the small remnant habitat has been largely destroyed by road construction and the surrounding monoculture of coffee plantations. Moreover, the remaining plants are threatened by flood or riverbank erosion. Given the small remnant habitat, in situ conservation seems pointless. However, reintroduction near the original habitat and assisted colonization of a new population at a higher elevation could be reasonable approaches to protect this species. Our recent work suggested that *P. spicerianum* can associate with a wide range of mycorrhizal fungi. The low specificity of adult plants for mycorrhizal partner suggests that the choice of locations to establish new populations might be broader than initially anticipated (Han et al. 2016). Based on seed storage and the development of an asymbiotic seed

Fig. 3 Transplanting seedlings of *Paphioepdilum spicerianum* at Tianzi Reserve

germination system, we obtained massive numbers of seedlings (Chen et al. 2015) and began reintroduction and assisted colonization in 2015 (Fig. 3).

2.2 *Implementation*

Following the idea of orchid integrated conservation, ex situ conservation of living plants, seed storage, developing a propagation system via in vitro seed germination, monitoring population dynamics, and studying pollination ecology and mycorrhizal fungi diversity were conducted from 2010. Thousands of seedlings were produced successfully and ready for reintroduction at the beginning of 2015. Based on the studies of pollination ecology and mycorrhizal fungi diversity, we believe that the choices of locations for new populations might be broader than initially thought, but for the long-term maintenance of new populations, lack of pollinators might be a great limitation. To fully establish a self-sustaining population of *P. spicerianum*, we suggested that some co-flowering plants should be introduced into assisted colonization sites with *P. spicerianum* at the same time.

Fig. 4 After 2 years at June 2017, the transplanted plants of *Paphioepdilum spicerianum* growing very well at Tianzi Reserve

2.3 Post-release Monitoring

The results of post-release monitoring showed that the survival ratio of reintroduction was much lower than assisted colonization (40% vs. 80%) after 2 years in the end of 2017, and only two plants remained in the reintroduction site in the end of 2018. Conversely, plants in the site of assisted colonization grow very well, and many plants flowered in 2018 and 2019 (Fig. 4). These results suggested that it is hard to establish a reintroduced population in its original habitat because it is seriously threatened by many factors, and assisted colonization could be a better way to save this species. However, it is uncertain if seeds could successfully germinate with compatible fungi in assisted colonization site, and this could be the major factor that limits potential for full recovery of a self-sustaining population. It is also important to choose appropriate and safe sites for reintroduction and assisted colonization, and it is necessary to conduct regular management after seedling transplantation.

3 Problems and Recommendations

- The original habitat is seriously threatened by many factors, and it is hard to establish a reintroduced population in original habitat.

- As a PSESP species, all seedlings come from just several mother plants, which may result in low genetic diversity of reintroduced and transplanted populations.
- For assisted colonization, i.e., a new population established outside of the original habitat, reproductive success may be limited by lack of effective pollinators, which could be the major factor that limits potential for full recovery of a self-sustaining population. It is uncertain now if seeds could successfully germinate with compatible fungi for introduced populations.
- Basic researches, such as seed storage, in vitro seed germination, monitoring of population dynamics, studies on pollination ecology and mycorrhizal fungi diversity, etc., are essential to conduct reintroduction or assisted colonization of endangered orchids.
- Sites selection is very important for future success of new population establishment, and regular management is necessary after seedling released.

Acknowledgments This chapter was modified from the paper published by our group in *Global Re-introduction Perspectives: 2018 Case Studies from Around the Globe edited by* Soorae, PS (Gao et al. 2018; pp. 247–250). The related contents are reused with permission.

References

Y. Chen, U.M. Goodalea, X.L. Fan, et al., Asymbiotic seed germination and *in vitro* seedling development of *Paphiopedilum spicerianum*: An orchid with an extremely small population in China. Glob. Ecol. Conserv. **3**, 367–378 (2015)

J.Y. Han, H.F. Xiao, J.Y. Gao, Seasonal dynamics of mycorrhizal fungi in *Paphiopedilum spicerianum* (Rchb. f) Pfitzer—A critically endangered orchid from China. Global Ecol. Conserv. **6**, 327–338 (2016)

H. Rankou, S. Molur. Paphiopedilum spicerianum. The IUCN Red List of Threatened Species 2015: e.T15052359A15055149. (2015). doi:https://doi.org/10.2305/IUCN.UK.2015-2.RLTS.T15052359A15055149.en

H. Ren, Q.M. Zhang, H.F. Lu, et al., Wild plant species with extremely small populations require conservation and reintroduction in China. Ambio **41**, 913–917 (2012)

State Forestry Administration of China, *The Saving and Conservation Program on Extremely Small Populations in China* (State Forestry Administration of China, Beijing, 2012)

D.P. Ye, Y.B. Luo, *Paphiopedilum spicerianum*, a new record of Orchidaceae from China. Acta Phytotax. Sin. **44**, 471–473 (2006)

Orchid Conservation Translocation Efforts in China

Hong Liu and Jiangyun Gao

Abstract The Global Strategy for Plant Conservation is a target-oriented plant conservation strategy. Among the targets was a call for 75% of threatened taxa to be included in ex situ collections and 20% to be available for recovery and reintroduction programs. However, the percentage of threatened taxa worldwide subject to reintroduction is still far less than this target. China is a recognized center of plant diversity globally. An estimated 11% of Chinese plants were threatened with extinction. The use of conservation translocation is relatively new in China, but quickly gaining popularity. In this chapter, we aimed to understand the extent of conservation translocation of Chinese orchids, a group of Chinese plants which were threatened with disproportionally high numbers, and the motivations behind these actions. We found that a total of 43 Chinese orchid species of 21 genera had been subject to conservation translocation and a total of 55 cases of translocations were recorded. Stakeholders who carried out these orchid restoration cases were motivated differently but could be grouped into the following three types. They were (1) rescue action by a nature reserve, (2) research experiments by research institutes and botanical gardens, and (3) sustainable use reintroduction by a private company. In the future, using semi-wild cultivation of high-value orchid species as a leverage to carry out species and forest restoration should be strongly encouraged. Considering the high demand for certain orchids, especially the persisting demand for wild-sourced plants, the semi-wild cultivation may be able to motivate more stakeholders

H. Liu (✉)
Center for Tropical Plant Conservation and Department of Earth and Environment, Florida International University, Miami, FL, USA

Fairchild Tropical Botanic Garden, Coral Gables, FL, USA
e-mail: hliu@fiu.edu

J. Gao
Yunnan Key Laboratory of Plant Reproductive Adaption and Evolutionary Ecology, Yunnan University, Kunming, Yunnan, China

Laboratory of Ecology and Evolutionary Biology, Yunnan University, Kunming, Yunnan, China
e-mail: jiangyun.gao@ynu.edu.cn

H. Ren (ed.), *Conservation and Reintroduction of Rare and Endangered Plants in China*, https://doi.org/10.1007/978-981-15-5301-1_19

to buy in the conservation of the species and their associated ecosystem, in China and other developing countries with strong plant use traditions.

Keywords Conservation motivation · Plant reintroduction · Plant conservation · Orchid reintroduction · Sustainable harvest · Sustainable use

1 Introduction

Globally, an estimated 22–62% of the total 310,000–422,000 seed plant species (Prance et al. 2000; Govaerts 2001) were threatened with extinction (Pitman and Jorgensen 2002). The global plant conservation community has currently adopted an integrated approach where conservation translocations—a generic term encompassing the movement of plants for a conservation purpose—is used increasingly as a tool to support species recovery and to supple the imperative in situ conservation (Maunder 1992; Havens et al. 2006). Documentation of efforts in conservation translocation (Godefroid et al. 2011; Maschinski and Haskins 2012; Liu et al. 2015; Reiter et al. 2016) and of supporting technical guidelines is increasing, e.g., Vallee et al. (2004), Guerrant and Kaye (2007), Menges (2008), Godefroid et al. (2011), Maschinski and Haskins (2012), and IUCN (2013).

The Global Strategy for Plant Conservation (CBD 2012) was written to provide target-oriented plant conservation strategy. Among the 16 targets was a call for 75% of threatened taxa to be included in ex situ collections and 20% to be available for recovery and reintroduction programs. However, the percentage of threatened taxa subject to reintroduction is still far less than this target. The total number of plant species subject to conservation translocation is estimated to be around 1050 (a sum of species reviewed in Dalrymple et al. (2012), Godefroid et al. (2011), Liu et al. (2015), and Reiter et al. (2016)), which accounted for about 0.3% of the world's total seed plants.

China is a recognized center of plant diversity globally, with 8–12% (35,000) of the world's higher plants (Zhang and Gilbert 2015; Qin et al. 2017) and an estimated 10,000 endemic plant species (Yang et al. 2005). Habitat loss and deterioration and overexploitation were the top factors driving the decline of Chinese species (Qin et al. 2017). Qin et al. (2017) considered 40 plant species extinct in China and further classified 3879 (10.8%) as threatened. It is apparent that the threats to plant diversity in China have been severe and are likely to keep increasing in the foreseeable future due to continued economic development and population growth (Qin et al. 2017; Ren et al. 2019). The use of conservation translocation is relatively new in China, but quickly gaining popularity. As of 2014, more than 150 plant species (3.8% of the Chinese threatened taxa) have been subjected to conservation translocation (Liu et al. 2015). More than 20% of these conservation translocation cases described in Liu et al. (2015) were orchids, a group of Chinese plants which were threatened with disproportionally high numbers (43%, Ministry of Environmental Protection of China 2013).

In this chapter, we aimed to understand the extent of conservation translocation of Chinese orchids and the motivations behind these actions. We asked the following questions: (1) which and how many orchid species were translocated for conservation purposes? (2) Were they threatened species? (3) What types of translocations were they (see the next section for detail classifications)? (4) What organizations carried out these translocations? (5) What were the funding sources? To do this we compiled information on the identity of the orchid species that have been subjected to conservation translocations and the organizations or persons who carried out these translocations, as well as their funding sources. We also offered an analysis on what future priorities would be.

2 Methods

We recognized two general classification systems describing the types of conservation translocations. The first classification is based on translocation recipient site characteristics, i.e., augmentation referring to site hosting an extant but shrunken population that can be bolstered through augmentation, reintroduction sensu stricto site that used to have a population, and introduction site that never hosted a population (IUCN 2013; Liu et al. 2015; Reiter et al. 2016). The other classification is based on the source of planting materials. If the plant materials were derived from ex situ propagation at botanical institutes, it is viewed as a conventional reintroduction (Reiter et al. 2016). In contrast, if the planting materials directly came from a wild population somewhere else, it is termed *salvage* or *rescue* (Reiter et al. 2016; Liu et al. 2012a, b).

To collect records of Chinese orchid conservation translocations, we searched both Chinese and English databases. Specifically, we used the literature search methods detailed in Liu et al. (2015). Since Liu et al. (2015) did a thorough job of locating all Chinese plant reintroduction cases reported in both Chinese and English databases before 2015, we included here all the orchid cases included by Liu et al. (2015). We further located new cases published after 2015 by using a query similar to Liu et al. (2015). The only change that we made was to replace "plant" with "orchid" in the search phrases to focus our search on orchids only. For the search in the Chinese scientific database, we used the following query: 兰花 (meaning orchid) AND 保育 (meaning conservation) AND 回归 (meaning reintroduction) OR 恢复 (meaning restoration). For search of the Chinese popular article on orchid translocation, we used the following query: 濒危兰花重引入或野外归放 (meaning endangered orchid reintroduction), 或 (meaning or), and 野生兰花迁移或抢救 (meaning wild orchid transplant or rescue).

We verified cases discussed in the newspaper articles, social media WeChat (a Chinese social media platform similar to Facebook in the USA), and government reports by contacting people or organizations identified in the reports or chat. We asked the same questions that were detailed in Liu et al. (2015). In addition, we reached out to Chinese colleagues whom we knew had worked on orchid conservation and requested copies of their conference abstracts, books or book chapters, and other published articles. We verified species names and endemism status using the online version of the Flora of China (FOC).

We used three project-level parameters to measure translocation success: percentage of transplants still alive at the time of survey, whether any plants in a translocation population flowered, and whether any plants in a translocation population fruited. We were able to gather data on factors likely to have some influence on the performance of the translocation populations: propagule type, propagule source, number of propagules, time since planting to monitoring, and conservation translocation type. There were four translocation types: augmentation, reintroduction within range, introduction, and conservation introduction. There were three propagule types: adult plants, saplings, and seeds. Propagule sources had three categories: same population propagated ex situ, rescued plants from nearby populations or mixed populations.

3 How Many and Which Chinese Orchids Were Subjected to Conservation Translocation?

A total of 43 Chinese orchid species of 21 genera were found to have been subject to conservation translocation of various types (Table 1.). Among these only two were Chinese endemic species and 19 were epiphytic. Twenty-five species were classified as threatened (five critically endangered, nine endangered, and ten vulnerable) by the Chinese Red List (Qin et al. 2017). This represented 3.7% of 666 threatened orchid species in China. This percentage is lower than that reported in Australia for Australian orchids, in which 12 threatened Australian orchid species, 6.5% of the 185 Australian threatened orchids, have been under a reintroduction program (Reiter et al. 2016). This percentage is likely the best among the world given that the majority of the handful of published orchid restoration studies were based on Australian orchids (e.g., Reiter et al. 2016). However, the total number of plant species subject to conservation translocation is estimated to be around 1050, based on data previously reported (Dalrymple et al. 2012; Godefroid et al. 2011), which accounted for about 0.3% of the world's total seed plants. By comparison, the percentage of threatened orchids that were subject to conservation translocation in China and Australia are likely higher than many other groups of plants.

A total of 55 cases of translocations were recorded as some species were translocated more than once (Table 2). The number of propagules used ranged from 2 to 10,000, with the extremely small numbers associated with rescue cases (Table 2.). While 31 translocations used plants rescued from a hydropower project (Liu et al. 2012a), in the remaining 24 translocation cases, planting materials were propagated ex situ, either from seeds collected in the wild by hand-pollination or from seeds produced by wild maternal plants raised in nurseries. There were 17 cases of augmentation, 11 reintroduction s.s., 20 introduction within range, and seven

Table 1 Chinese orchids that were subject to conservation translocation, their habit, and conservation status assessment

Species	Chinese name	Endemic	Habit	Red List Global (IUCN)	Red List Chinese	Note
Acanthephippium sylhetense Lindley	坛花兰	No	Terrestrial	Not assessed	VU	Threatened
Calanthe argenteostriata C. Z. Tang & S. J. C	银带虾脊兰	No	Terrestrial	Not assessed	LC	Non-threatened
Changnienia amoena S. S. Chien	独花兰	Yes	Terrestrial	EN	EN	Threatened
Cheirostylis chinensis Rolfe	中华叉柱兰	No	Epiphyte	Not assessed	LC	Non-threatened
Cleisostoma nangongense Z. H. Tsi	南贡隔距兰	Yes	Epiphyte	Not assessed	VU	Threatened
Cleisostoma paniculatum (Ker Gawler) Garay	大序隔距兰	No	Epiphyte	Not assessed	LC	Non-threatened
Cleisostoma williamsonii (Rchb. f.) Garay	红花隔距兰	No	Epiphyte	Not assessed	LC	Non-threatened
Cymbidium aloifolium (Linnaeus) Swartz	纹瓣兰	No	Terrestrial	Not assessed	NT	Non-threatened
Cymbidium mannii H. G. Reichenbach	硬叶兰	No	Terrestrial	Not assessed	NT	Non-threatened
Cymbidium goeringii (H. G. Reichenbach) H. G. Reichenbach	春兰	No	Terrestrial	Not assessed	VU	Threatened
Cymbidium tortisepalum var. *longibracteatum*	春剑	Yes	Terrestrial	Not assessed	EN	Threatened
Cymbidium tracyanum L. castle	西藏虎头兰	No	Epiphyte	LC	LC	Non-threatened
Dendrobium catenatum	黄石斛/铁皮石斛	No	Epiphyte	Not assessed	CR	Threatened
Dendrobium cucullatum R. Brown	兜唇石斛	No	Terrestrial	Not assessed	VU	Threatened
Dendrobium denneanum Kerr	叠鞘石斛	No	Epiphyte	Not assessed	VU	Threatened
Dendrobium lindleyi Steudel	聚石斛	No	Epiphyte	Not assessed	LC	Non-threatened
Dendrobium loddigesii Rolfe	美花石斛	No	Epiphyte	Not assessed	VU	Threatened
Dendrobium nobile Lindl.	石斛/金钗石斛	No	Epiphyte	Not assessed	VU	Threatened

(continued)

Table 1 (continued)

Species	Chinese name	Endemic	Habit	Red List		Note
				Global (IUCN)	Chinese	
Dendrobium williamsonii J. Day & H. G. Reich	黑毛石斛	No	Epiphyte	EN	EN	Threatened
Dendrobium devonianum Paxton	齿瓣石斛	No	Epiphyte	EN	EN	Threatened
Eria corneri H. G. Reichenbach	半柱毛兰	No	Terrestrial	Not assessed	LC	Non-threatened
Eulophia bracteosa Lindley	长苞美冠兰	No	Terrestrial	Not assessed	VU	Threatened
Geodorum eulophioides Schlechter	贵州地宝兰	No	Terrestrial	Not assessed	EN	Threatened
Geodorum recurvum (Roxburgh) Alston	多花地宝兰	No	Terrestrial	Not threatened	NT	Threatened
Habenaria dentata (Swartz) Schlechter	鹅毛玉凤花	No	Terrestrial	Not assessed	LC	Non-threatened
Liparis viridiflora (Blume) Lindley	长茎羊耳蒜	No	Terrestrial	Not assessed	LC	Non-threatened
Luisia teres (Thunberg) Blume	叉唇钗子股	No	Epiphyte	Not assessed	NT	Non-threatened
Oberonia cavaleriei Finet	棒叶鸢尾兰	No	Epiphyte	Not assessed	LC	Non-threatened
Paphiopedilum armeniacum S. C. Chen & F.	杏黄兜兰	No	Terrestrial	EN	CR	Threatened
Paphiopedilum dianthum Tang & F. T. Wang	长瓣兜兰	No	Terrestrial	EN	VU	Threatened
Paphiopedilum emersonii Koop. & P. J.Cribb	白花兜兰	No	Terrestrial	CR	CR	Threatened
Paphiopedilum hirsutissimum Lindley ex Hook	带叶兜兰	No	Terrestrial	VU	VU	Threatened
Paphiopedilum malipoense S. C. Chen & Z. H.	麻栗坡兜兰	No	Terrestrial	EN	CR	Threatened
Paphiopedilum micranthum Tang & F. T. Wan	硬叶兜兰	No	Terrestrial	CR	EN	Threatened
Paphiopedilum purpuratum (Lindl.) Stein	紫纹兜兰	No	Terrestrial	CR	EN	Threatened
Paphiopedilum spicerianum Pfitzer	白旗兜兰	No	Terrestrial	EN	EN	Threatened

(continued)

Table 1 (continued)

Species	Chinese name	Endemic	Habit	Red List		Note
				Global (IUCN)	Chinese	
Paphiopedilum wardii Summerhayes	彩云兜兰	No	Terrestrial	EN	DD	Threatened
Phaius flavus (Blume) Lindley	黄花鹤顶兰	No	Terrestrial	Not assessed	LC	Non-threatened
Pholidota yunnanensis Rolfe	云南石仙桃	No	Epiphyte	Not assessed	NT	Non-threatened
Renanthera imschootiana Rolfe	云南火焰兰	No	Epiphyte	Not assessed	CR	Threatened
Rhynchostylis retusa (Linnaeus) Blume	钻喙兰	No	Epiphyte	Not assessed	EN	Threatened
Robiquetia succisa (Lindley) Seidenfaden & Garay	大叶寄树兰	No	Epiphyte	Not assessed	LC	Non-threatened
Vandopsis gigantea (Lindley) Pfitzer	拟万代兰	No	Epiphyte	Not assessed	LC	Non-threatened

introduction out of range. Among these, 31 cases had post-translocation monitoring data with survival and reproduction information recorded, while the remaining 24 cases were either not monitored or were too new to have survival or reproductive data. The last monitoring time since planting ranged from 0.5 to 14 years, with survival rates ranging from 0% to 100%. Fourteen translocation cases used saplings only, 31 cases used adult plants only, one case used seeds, and the remaining eight used a mixture of saplings and adult plants. In 33 cases the transplants produced flowers, and among these, six cases reported fruit production. Seven cases did not record flowers, and the remaining 15 cases did not have a record on this variable (missing data).

4 Motivations of the Chinese Orchid Conservation Translocation: The Who and Why Questions

Stakeholders who carried out these orchid restoration cases were motivated differently but could be grouped into the following three types. They were (1) rescue action by a nature reserve, (2) research experiments by research institutes and botanical gardens, and (3) sustainable use reintroduction by a private company. Note that some research questions also centered on sustainable use.

Table 2 Chinese orchid reintroduction locations, types, and monitoring information

Species	Reintroduction information			Propagule information			Time to monitor (year)	Survival (%)	Reproduction	Implementation agency	Fund. sources
	Location	Type	Date	Size	Number	Source					
Acanthephippium sylhetense	Yachang NONR, Guangxi	Intro_w	2006	Adult	3	Nearby pop_r	5	25	Flowered	YNNR	DCC company
Calanthe argenteo-striata	Yachang NONR, Guangxi	Intro_o	2006	Adult	27	Nearby pop_r	5	93.50	Flowered	YNNR	DCC company
Cheirostylis chinensis	Yachang NONR, Guangxi	Intro_w	2006	Adult	5	Nearby pop_r	5	100	Flowered	YNNR	DCC company
Cleisostoma nangongense	Yachang NONR, Guangxi	Intro_w	2006	Adult	17	Nearby pop_r	5	59	Flowered	YNNR	DCC company
Cleisostoma paniculatum	Yachang NONR, Guangxi	Intro_w	2006	Adult	3	Nearby pop_r	5	100	Flowered	YNNR	DCC company
Cymbidium aloifolium	Yachang NONR, Guangxi	Intro_w	2006	Adult	14	Nearby pop_r	5	87.90	Flowered	YNNR	DCC company
Cymbidium mannii	Yachang NONR, Guangxi	Intro_o	2006	Adult	55	Nearby pop_r	5	87.90	Flowered	YNNR	DCC company
Cymbidium mannii	Xishuangbanna, Yunnan	Augmt	*2012*	*Saplings*	*480*	Same pop_e	2	*1.6*	Not flowered	*XTBG*	NKRDPC
Cymbidium mannii	Xishuangbanna, Yunnan	Augmt	*2012–2014*	*Saplings*	*480*	Same pop_e	*0.5*	*93.5*	Not flowered	*XTBG*	NKRDPC
Cymbidium tortisepalum. var. *longibracteatum*	Yachang NONR, Guangxi	Intro_w	2006	Adult	3	Nearby pop_r	5	24.10	Flowered	YNNR	DCC company
Cymbidium tracyanum	Yachang NONR, Guangxi	Intro_o	2006	Adult	24	Nearby pop_r	5	77.10	Flowered	YNNR	DCC company
Dendrobium lindleyi	Yachang NONR, Guangxi	Intro_o	2006	Adult	26	Nearby pop_r	5	100	Flowered	YNNR	DCC company
Dendrobium williamsonii	Yachang NONR, Guangxi	Intro_w	2006	Adult	2	Nearby pop_r	5	33.30	Flowered	YNNR	DCC company
Dendrobium devonianum	Xishuangbanna, Yunnan	Augmt	*2014*	*Seeds with fungi*	*>5000*	Same pop_e	1	*0.24*	Not flowered	*XTBG*	NKRDPC
Eria corneri	Yachang NONR, Guangxi	Augmt	2006	Adult	29	Nearby pop_r	5	66.70	Flowered	YNNR	DCC company
Liparis viridiflora	Yachang NONR, Guangxi	Intro_w	2006	Adult	2	Nearby pop_r	5	100	Flowered	YNNR	DCC company

Luisia teres	Yachang NONR, Guangxi	Augmt	2006	Adult	2	Nearby pop_r	5	40	Flowered	YNNR	DCC company
Paphiopedilum armeniacum	Luoshanpo, Yunnan	Augmt	2003	Saplings_ t	160	Same pop_e	2	38.13	Flowered and fruited	SZNOG	NP-WLCNRC
Paphiopedilum armeniacum	Gulinqing, Yunnan	Reintro	2009	Saplings	600	Mixed pop_e	5	80	Flowered and fruited	KIB	BGCI China
Paphiopedilum armeniacum	Gaoligong, Yunnan	Reintro	2009	Saplings	600	Mixed pop_e	5	80	Flowered and fruited	KIB	BGCI China
Paphiopedilum dianthum	Yachang NONR, Guangxi	Augmt	2006	Adult	21	Nearby pop_r	5	100	Flowered	YNNR	DCC company
Paphiopedilum hirsutissimum	Yachang NONR, Guangxi	Augmt	2006	Adult	22	Nearby pop_r	5	73.30	Flowered	YNNR	DCC company
Paphiopedilum malipoense	Malipo, Wenshan, Yunnan	Augmt	2004	Saplings_ t	5000	Same pop_e	14	99	Flowered, fruited, recruited	SZNOG	NP-WLCNRC
Paphiopedilum micranthum	Yachang NONR, Guangxi	Augmt	2006	Adult	2	Nearby pop_r	5	100	Flowered	YNNR	DCC company
Paphiopedilum spicerianum	Pu'er, Yunnan	Augmt	2015	Saplings	30	Same pop_e	2	40	Not flowered	*YNU*	NKRDPC
Paphiopedilum spicerianum	Xishuangbanna	Intro_o	2015	Saplings	30	Same pop_e	2	80	Not flowered	*YNU*	NKRDPC
Paphiopedilum wardii	Yachang NONR, Guangxi	Intro_w	2011	Saplings_ t	100	Nearby pop_r	2	75	Flowered and fruited	SCBG	NSFC
Phaius flavus	Yachang NONR, Guangxi	Intro_w	2006	Adult	1	Nearby pop_r	5	100	Flowered	YNNR	DCC company
Rhynchostylis retusa	Xishuangbanna, Yunnan	Augmt	*2012*	*Saplings*	*480*	Same pop_e	2	*4.8*	Not flowered	*XTBG*	NKRDPC
Rhynchostylis retusa	Xishuangbanna, Yunnan	Augmt	*2012*	*Saplings*	*480*	Same pop_e	*0.5*	*100*	Not flowered	*XTBG*	NKRDPC
Vandopsis gigantea	Yachang NONR, Guangxi	Intro_w	2006	Adult	34	Nearby pop_r	5	0	Flowered	YNNR	DCC company
Cases with no monitoring or incomplete data											
Changienia amoena	Jiugongshan, Hubei	Augmt	1980	Adults	–	–	–	–	Flowered, fruited, recruited	WBG	NSFC
Cleisostoma williamsonii	Yachang NONR, Guangxi	Intro_w	2006	Adults	5	Nearby pop_r	–	–	Flowered	YNNR	DCC company

(continued)

Table 2 (continued)

Species	Reintroduction information			Propagule information			Time to monitor (year)	Survival (%)	Reproduction	Implementation agency	Fund. sources
	Location	Type	Date	Size	Number	Source					
Cymbidium goeringii	Yachang NONR, Guangxi	Intro_w	2006	Adults	13	Nearby pop_r	–	–	–	YNNR	DCC company
Dendrobium cucullatum	Yachang NONR, Guangxi	Intro_w	2006	Adults	6	Nearby pop_r	–	–	Flowered	YNNR	DCC company
Dendrobium denneanum	Yachang NONR, Guangxi	Intro_w	2006	Adults	5	Nearby pop_r	–	–	Flowered	YNNR	DCC company
Dendrobium catenatum	Xingyi, Guizhou	Reintro	2019	Saplings and adults	10,000	Mixed_e	–	–	–	Lvyuan	Self
Dendrobium fimbriatum	Xingyi, Guizhou	Reintro	2019	Saplings and adults	–	Mixed_e	–	–	–	Lvyuan	Self
Dendrobium loddigesii	Yachang NONR, Guangxi	Intro_w	2006	Adults	15	Nearby pop_r	–	–	–	YNNR	DCC company
Dendrobium nobile	Xingyi, Guizhou	Reintro	2019	Saplings and adults	10,000	Mixed_e	–	–	–	Lvyuan	Self
Eulophia bracteosa	Yachang NONR, Guangxi	Intro_w	2006	Adults	20	Nearby pop_r	–	–	Flowered	YNNR	DCC company
Geodorum eulophioides	Yachang NONR, Guangxi	Intro_w	2006	Adults	–	Nearby pop_r	–	–	–	–	–
Geodorum recurvum	Yachang NONR, Guangxi	Intro_w	2006	Adults	12	Nearby pop_r	–	–	–	YNNR	DCC company
Habenaria dentata	Yachang NONR, Guangxi	Intro_w	2006	Adults	115	Nearby pop_r	–	–	Flowered	YNNR	DCC company
Oberonia myosurus	Yachang NONR, Guangxi	Augmt	2006	Adults	21	Nearby pop_r	–	–	–	YNNR	DCC company
Paphiopedilum armeniacum	Xingyi, Guizhou	Reintro	2019	Saplings and adults	500	Same pop_e	–	–	–	Lvyuan	Self
Paphiopedilum dianthum	Xingyi, Guizhou	Reintro	2019	Saplings and adults	500	Same pop_e	–	–	–	Lvyuan	Self
Paphiopedilum emersonii	Xingyi, Guizhou	Reintro	2019	Saplings and adults	500	Same pop_e	–	–	–	Lvyuan	Self

Paphiopedilum hirsutissimum	Xingyi, Guizhou	Reintro	2019	Saplings and adults	500	Same pop_e	–	–	–	Lvyuan	Self
Paphiopedilum malipoense	Xingyi, Guizhou	Reintro	2019	Saplings and adults	500	Same pop_e	–	–	–	Lvyuan	Self
Paphiopedilum malipoense	Malipo, Wenshan, Yunnan	Augmt	2004	Saplings_ t	5000	Same pop_e	14	99	Flowered, fruited, recruited	SZNO	NP-WLCNRC
Paphiopedilum micranthum	Xingyi, Guizhou	Reintro	2019	Saplings and adults	500	Same pop_e	–	–	–	Lvyuan	Self
Paphiopedilum purpuratum	Shenzhen, Guangdong	Reintro	2018	Adults	500	Nearby pop_e	–	–	–	SZNO	NP-WLCNRC
Pholidota yunnanensis	Yachang NONR, Guangxi	Augmt	2006	Adults	4	Nearby pop_r	–	–	–	YNNR	DCC company
Renanthera imschootiana	Yuanjiang NR, Yunnan	Augmt	2009	6-m and 1-year saplings	600	Same pop_e	2	65	Flowered	SCBG	PG
Renanthera imschootiana	Yuanjiang NR, Ehuangzhuang, NR, Guangdong	Intro_o	2009	6-m and 1-year saplings	600	Same pop_e	2	77	Flowered	SCBG	PG
Renanthera imschootiana	Huolu Mountain Forest park, Guangdong	Intro_o	2009	6-m and 1-year saplings	600	Same pop_e	2	62	Flowered	SCBG	PG

Reintroduction type: *reintro* reintroduction s.s., *augmt* augmentation, *intro_w* within range introduction, *intro_o*, out of range introduction

Propagule sources: *same pop_e* propagated ex situ using materials from the same site, *nearby pop_r* rescued from a nearby population, *nearby pop_e* propagated ex situ using material from a nearby population, *mixed_e* propagated ex situ using mixed materials from same and nearby populations

Implementation agency: *SCBG* Southern China Botanical Garden, *KIB* Kunming Institute of Botany, *YIES* Yunnan Institute of Environmental Science, *SZNO* Shenzhen National Orchid Germplasm Resources Conservation Center, *LQVEI* Laboratory of Quantitative Vegetation Ecology, Institute of Botany, the Chinese Academy of Sciences, *HAASC* Hunan Academy of Agricultural Sciences of China, *NRI* National Research Institute, *SCAU* South China Agricultural University, *YAU* Yunnan Agricultural University, *WBG* Wuhan Botanical Garden, *YNNR* Yachang Orchid National Nature Reserve, *YNU* Yunnan University, Lvyuan, Guizhou Qiandongnan Prefecture Lvyuan Animals and Plants Technological Development Co. Ltd.

Funding sources: *BGCI* Botanic Gardens Conservation International, *CSFA* China State Forestry Administration, *NKRDPC* the National Key R and D Program of China, *NP-WLCNRC* National Program for Wildlife Conservation and Nature Reserve Construction, *NSFC* National Natural Science Foundation of China, *PG* provincial or other local government, *Self* self-funded project.

4.1 Rescue Cases

The rescue conservation translocations were carried out by the Yachang National Nature Reserve in Guangxi Zhuang Autonomous Region, hereafter referred to as Yachang Reserve, in response to damming of a river (Fig. 1a, Liu et al. 2012a, b). It occurred in 2006 in and near the Yachang Reserve, a remote area in Southwestern China. The Yachang Reserve was the only Chinese nature reserve that was established to protect wild orchids specifically (Fig. 1b). This remote and relatively small (220 km^2) nature reserve is home to more than 150 species of orchids, some of

Fig. 1 A stretch of the Hongshui River near the Yachang National Orchid Nature Reserve in Guangxi Zhuang Autonomous Region, showing the high watermarks when the dam was in full operation (**a**), and adjacent areas in Southwestern China consist of numerous hills like these, no more than 1200 m above sea level, and usually rich in orchids (**b**) (Photo credit: Hong Liu)

them in extremely large, relatively undisturbed populations (Liu et al. 2010). The Hongshui River, one of the two major rivers flowing through or adjacent to the Yachang Reserve, is the site of one of China's largest hydropower projects (Fig. 1a). When the project was near completion, an unprecedented wild plant rescue mission, funded by the hydroelectric power project construction company, took place (Liu et al. 2012a). Nearly 1000 delicate, reproductive-size orchid plants from 29 species and 16 genera from locations slated for flooding were carefully translocated (Fig. 2). The Yachang Reserve staff relocated these orchids to a fenced, patrolled forest site at approximately 1000 m above sea level—as much as 600 m higher than their original locations, due to poaching concerns (Liu et al. 2012a, b).

Even though less than half (41.4%) of the species involved in the rescue were classified as non-threatened, the rescue is still of conservation significance. It also has research significance. For more than half of the orchid species, the relocation placed them out of their natural elevation range, which offered a unique opportunity to test the new conservation idea of assisted colonization (i.e., conservation introduction). The fates of many translocated plants were not followed, but many others were (Liu et al. 2012b). A few species managed to flower, fruit, and recruit. Orchids that were subject to either conservation introduction or conventional translocation were able to establish new relationships with root fungi, including known orchid mycorrhizal fungi (OMF) groups (Downing et al. 2017). This research provides the first examination on how critical biotic interactions change following conservation introduction compared to that of conventional translocation. Researches like Downing et al. (2017) are important in evaluating the feasibility of conservation introduction.

4.2 Conservation Research Cases

The conservation research cases were carried out by botanical gardens with conservation research agenda (e.g., Liu et al. 2006; Gao et al. 2020). The main group of species that fell into this category was the lady's slipper orchids (Fig. 3), in the genus *Paphiopedilum*, carried out by three national botanical institutes: Xishuangbanna Tropical Botanical Garden (Gao et al. 2020), South China Botanical Garden (Zeng et al. 2012), and the Shenzhen National Orchid Germplasm Resources Conservation Center (Liu et al. 2006). The lady's slipper orchids are a group subject to heavy collecting pressure because of illegal international trade (Chen 1996, 1999; Luo et al. 2003a, b).

Several other conservation efforts were led by a research group at the XTBG, who carried out important experimental reintroductions to answer various research questions. One experiment tested the impact of forest fragmentation on reintroduction success, using two orchid species, i.e., *Cymbidium mannii* and *Rhynchostylis retusa*; the latter is a threatened species. The experiment was carried out at two sites, one in a limestone forest fragment surrounded by rubber plantations and the other within a

Fig. 2 A sample of the 29 rescued orchid species from the flooding area near Yachang National Orchid Nature Reserve in Guangxi Zhuang Autonomous Region include *Geodorum eulophioides* (**a**), *Paphiopedilum dianthum* (**b**), *Luisia teres* (**c**), *Cymbidium tracyanum* (**d**), and *Dendrobium loddigesii* (**e**) (Photo credit: Hong Liu)

Fig. 3 *Paphiopedilum purpuratum* is one of the lady's slipper orchids that were translocated by the Shenzhen National Orchid Germplasm Resources Conservation Center, a state research institute focusing on research and ex situ collection of Chinese orchids (Photo credit: Hong Liu)

nature reserve. They found that survival of orchids planted in fragmented forest areas during the dry season was the lowest.

Another experiment involved the reintroduction of *Dendrobium devonianum*, a species harvested for medicinal use by locals, in tea plantations. The researchers found that by using the appropriate fungal species, the seeds can germinate in situ. This experiment resulted in a restored population and established a technique for locals to reintroduce the depleted populations on their own.

4.3 Sustainable Use Reintroductions and Augmentations

The sustainable use reintroductions were carried out by a large private company located in an area of high orchid diversity, Xingyi County, Guizhou. Wild orchids were traded for both ornamental and medicinal purposes. Modern artificial propagation techniques, though existing for several decades, were not adopted in China until about a decade ago and only by a handful of large companies. Trading of wild collected orchids is common (Gale et al. 2019; Wong and Liu 2019), largely because it is cheaper to collect from the wild than to use artificial cultivation (Gale et al. 2019). Depletion of wild orchid populations was reported in formerly orchid-rich areas of China, including Southwest China provinces such as Guangxi and Guizhou and the Xingyi city in southeast of Guizhou Province (Luo et al. 2013). Xingyi is a limestone dominated mountainous area famous for its abundance of orchids, especially species in the *Dendrobium* and *Paphiopedilum* genera. It used to be a traditional trading hub for wild-harvested dendrobiums used in Chinese medicine (Luo et al. 2013). The Lvyuan Animals and Plants Technological Development Co. Ltd. is the largest horticultural company with large-scale artificial propagation facilities in Xingyi and one of the largest orchid nurseries in China. It has access to

Fig. 4 A landscape with numerous limestone mountains in Xingyi, Guizhou Province, where reintroduction s.s. and augmentation of many native orchid species were carried out by a large private nursery, the Qiandongnan Autonomous Prefecture Lvyuan Animals and Plants Technological Development Co. Ltd. (Photo credit: Hong Liu)

national markets via the Internet and large national floral exhibits. Yet it faces steep competition from wild-sourced orchid trade in both physical and Internet platforms (Personal communication, Keyun Deng, owner and CEO of Lvyuan Co.). As a way to deal with the competition, the company embarked on reintroduction of eight orchid species native to nearby privately managed mountains (Table 2, Figs. 4–6). Populations of these orchids were extirpated due to unregulated collection. The goal of such reintroduction is to support sustainable use (Personal communication, Keyun Deng, owner and CEO of Lvyuan Company), meaning sustainable harvest in the near future is expected.

4.4 Future Priorities

In the future, using semi-wild cultivation of high-value and highly exploited orchid species as a leverage to carry out species and forest restoration should be strongly encouraged. This action should involve research institutes to determine the appropriateness of the species and source materials. For example, the expansion of the

Fig. 5 Selected plants of *Dendrobium catenatum* (**a**) and *Paphiopedilum malipoense* (**b**) transplanted in a limestone mountain in Xingyi, Guizhou Province, PR China (Photo credit: Hong Liu (**a**) and Keyun Deng (**b**))

genetic diversity of the orchids transplanted to the local mountains by Lvyuan Co. is not known. In addition, genetic contamination is a big concern during reintroduction, especially when it is done by private nurseries with profit as a primary goal. With appropriate partnership, both genetic contamination and bottleneck can be minimized (Vovides et al. 2010). In addition, research goals should also include determination of regimes for sustainable harvest of the various orchid species. One example of this involves *Dendrobium catenatum* (aka *D. officinale*, Fig. 5), an orchid used as a traditional Chinese medicine. Wild populations of *D. officinale* have been collected to near extinction and restoration is needed. Researchers are exploring the ecological and socioeconomic conditions under which farmers are motivated to reintroduce this species in restored and/or natural forests to achieve sustainable use of the species (Ticktin et al. 2020). This priority may be different from developed countries like the USA and Australia, where over collecting is a conservation concern but not a major one. For example, none of the orchids subject to reintroduction in Australia were subject to the trade of wild-collected plants (Reiter et al. 2016). Considering the high demand for certain orchids, especially the persisting demand for wild-sourced plants, either for medicinal purposes due to the belief that wild is better (Liu et al. 2019) or due to cost, i.e., wild is cheaper, the semi-wild cultivation of exploited plants may be able to motivate more stakeholders to buy in for the conservation of the species and their associated ecosystem (Liu et al. 2019; Ticktin et al. 2020).

Fig. 6 Selected lady's slipper orchids, *Paphiopedilum micranthum* (**a**), *P. malipoense* (**b**), and *P. armeniacum* (**c**) propagated in the nursery of the Qiandongnan Autonomous Prefecture Lvyuan Animals and Plants Technological Development Co. Ltd. in Xingyi, Guizhou Province. Some populations of these orchid species have been extirpated in the region and are targeted for reintroduction by the company (Photo credit: Keyun Deng)

References

CBD, *Global Strategy for Plant Conservation: 2011–2020* (Botanic Gardens Conservation International, Richmond, 2012)

S.C. Chen. Orchids and their conservation in China. Proc 5th Asia Pacific Orch Conf & Show; 1996; 49–52

S.C. Chen, *Paphiopedilum. Flora Reipublicye Popularis Sinicae*, vol 17 (Science Press, Beijing, 1999;. Tomus), pp. 52–72

S.E. Dalrymple, E. Banks, G.B. Stewart, et al., A meta-analysis of threatened plant reintroductions from across the globe, in *Plant Reintroduction in a Changing Climate: Promises and Perils*, ed. by J. Maschinski, K. E. Haskins, (Island Press, Washington, DC, 2012), pp. 31–50

J.L. Downing, H. Liu, S.C. Shao, X.L. Wang, M. McCormick, R.Y. Den, J.Y. Gao, Contrasting changes in biotic interactions of orchid populations subject to conservation introduction vs. conventional translocation in tropical China. Biol. Conserv. **212**, 29–38 (2017)

W.S. Gale, P. Kumar, A. Hinsley, et al., Quantifying the trade in wild-collected ornamental orchids in South China: Diversity, volume and value gradients underscore the primacy of supply. Biol. Conserv. **238** (2019). https://doi.org/10.1016/j.biocon.2019.108204

Gao JY, Fan XL, Yang WK. Reintroduction and Assisted Colonization of *Paphiopedilum spicerianum*: An Orchid with Extremely Small Population in China. Chapter 18, this volume. (2020)

S. Godefroid, C. Piazza, G. Rossi, et al., How successful are plant species reintroductions? Biol. Conserv. **144**, 672–682 (2011)

R. Govaerts, How many species of seed plants are there? Taxon **50**, 1085–1090 (2001)

E.O.J. Guerrant, T.N. Kaye, Reintroduction of rare and endangered plants: Common factors, questions and approaches. Austral J Bot. **55**, 362–370 (2007)

K. Havens, P. Vitt, M. Maunder, et al., *Ex situ* plant conservation and beyond. Bioscience **56**, 525–531 (2006)

IUCN, *Guidelines for Reintroductions and Other Conservation Translocations. Version1.0* (IUCN, Gland, 2013)

H. Liu, C.L. Feng, S.W. Cai, et al., Rescuing endangered orchids from a man-made flood. Trop. Garden Mag., 32–36 (2012a). Spring

H. Liu, C.L. Feng, B.S. Chen, et al., Overcoming extreme weather challenges: Successful but variable assisted colonization of wild orchids in southwestern China. Biol. Conserv. **150**, 68–75 (2012b)

H. Liu, C.L. Feng, Y.B. Luo, et al., Potential challenges of climate change to orchid conservation in a wild orchid hotspot in southwestern China. Bot. Rev. **76**, 174–192 (2010)

H. Liu, S.W. Gale, M.L. Cheuk, et al., Conservation impacts of commercial cultivation of endangered and overharvested plants. Conserv. Biol. **33**, 288–299 (2019)

H. Liu, H. Ren, Q. Liu, et al., Translocation of threatened plants as a conservation measure in China. Conserv. Biol. **29**, 1537–1551 (2015)

Z.J. Liu, K.W. Liu, L.J. Chen, et al., Conservation ecology of endangered species *Paphiopedilum armeniacum* (Orchidaceae). Acta Ecol. Sin. **26**, 2791–2800 (2006)

X.Q. Luo, M.K. Wu, G. Shen, et al., *Dendrobium officinale* reintroduction conservation and sustainable utilization in Guizhou Karst Areas. Chin. Wild Plant Resour. **32**, 47–50 (2013)

Y.B. Luo, J.S. Jia, C.L. Wang, A general review of the conservation status of Chinese orchids. Biodivers. Sci. **11**, 70–77 (2003a)

Y.B. Luo, J.S. Jia, C.L. Wang, Conservation strategy and potential advantages of the Chinese *Paphiopedilum*. Biodivers. Sci. **11**, 491–498 (2003b)

J. Maschinski, K.E. Haskins, *Plant Reintroduction in a Changing Climate: Promises and Perils* (Island Press, Washington D.C., 2012)

M. Maunder, Plant reintroduction: An overview. Biodivers. Conserv. **1**, 51–61 (1992)

E.S. Menges, Restoration demography and genetics of plants: When is a translocation successful? Aust. J. Bot. **56**, 187–196 (2008)

Ministry of Environmental Protection (MEP) of the People's Republic of China and the Chinese Academy of Sciences, *The Evaluation Report of the Chinese Biodiversity Red List—Higher Plants Volume* (MEP, Beijing. Available from http://www.mee.gov.cn/gkml/hbb/bgg/201309/W020130912562095920726.pdf (accessed Oct 2019), 2013)

N.C.A. Pitman, P.M. Jorgensen, Estimating the size of the world's threatened flora. Science **298**, 989 (2002)

G.T. Prance, H. Beentje, J. Dransfield, et al., The tropical flora remains undercollected. Ann Missouri Bot. Gard. **87**, 67–71 (2000)

H.N. Qin, L.N. Zhao, S.X. Yu, H.Y. Liu, B. Liu, N.H. Xia, et al., Evaluating the endangerment status of China's angiosperms through the red list assessment. Biodivers. Sci. **25**, 745–757 (2017). In Chinese with English abstract

N. Reiter, J. Whitfield, G. Pollard, et al., Orchid re-introduction: An evaluation of success and ecological considerations using key comparative studies from Australia. Plant Ecol. **217**, 81–95 (2016)

H. Ren, H.N. Qin, Z.Y. Ouyang, et al., Progress of implementation on the global strategy for plant conservation in (2011–2020) China. Biol. Conserv. **230**, 169–178 (2019)

T. Ticktin, D. Mondragón, L. Lopez-Toledo, et al., Synthesis of wild orchid trade and demography provides new insight on conservation strategies. Conserv. Lett., 1–10 (2020)

L. Vallee, T. Hogbin, L. Monks, et al., *Guidelines for the Translocation of Threatened Plants in Australia* (Canberra, Australian Network for Plant Conservation, 2004)

A.P. Vovides, M.A. Perez-Farrera, C. Iglesias, Cycad propagation by rural nurseries in Mexico as an alternative conservation strategy: 20 years on. Kew Bull. **65**, 603–611 (2010)

S. Wong, H. Liu, Wild-orchid trade in a Chinese e-commerce market. Econ. Bot. (2019). https://doi.org/10.1007/s12231-019-09463-2

Q.E. Yang, G.H. Zhu, D.Y. Hong, et al., World's largest flora completed. Science **309**, 2163 (2005)

S.J. Zeng, K.L. Wu, J.A. Teixeira da Silva, et al., Asymbiotic seed germination, seedling development and reintroduction of *Paphiopedilum wardii* Sumerh. An endangered terrestrial orchid. Sci. Hortic. **138**, 198–209 (2012)

L.B. Zhang, M.G. Gilbert, Comparison of classification of vascular plants of China. Taxon **64**, 17–26 (2015)

Reintroduction of Wild *Cycas* Species in China

Shuguang Jian, Xun Gong, and Nian Liu

Abstract There are more than 20 wild cycad species, with varied habitats in China. Most of the *Cycas* species in China have few populations and small population size. All wild species of *Cycas* are classified under State Protection Category I in China. Illegal excavation and trade and habitat loss are the major endangered factors of existing cycads in China, especially for those with high ornamental value. The Chinese government issued a series of policies and carried out a number of key programs for wild cycad species protection and reintroduction, which played a significant role in conservation of *Cycas* species in China. To date, there are only four *Cycas* species (*C. debaoensis, C. fairylakea, C. panzhihuaensis, C. diannanensis*) that have been reintroduced to the wild in China. Among these *Cycas* species, the reintroduction of *Cycas debaoensis* is rather successful. We investigated the distribution, population size, population ecology, reproductive biology (including seed germination), and population genetics of these *Cycas* species before reintroduction. After transplantation, we monitored the survival rate, height, and crown of transplants.

Keywords Wild *Cycas* species in China · Reintroduction · *C. debaoensis* · *C. fairylakea* · *C. panzhihuaensis* · *C. diannanensis*

S. Jian (✉)
South China Botanical Garden, Chinese Academy of Sciences, Guangzhou, China
e-mail: jiansg@scbg.ac.cn

X. Gong
Chinese Academy of Sciences, Kunming Institute of Botany, Kunming, Yunnan, China
e-mail: gongxun@mail.kib.ac.cn

N. Liu
Zhongkai University of Agriculture and Engineering, Guangzhou, China
e-mail: liunan@scbg.ac.cn

H. Ren (ed.), *Conservation and Reintroduction of Rare and Endangered Plants in China*, https://doi.org/10.1007/978-981-15-5301-1_20

1 Introduction

In China, there are more than 20 wild cycad species, distributed in the tropical and subtropical areas, with varied habitats. South China is viewed as the origin of *Cycas*. Most of the *Cycas* species in China have few populations and small population size, but have good value in scientific research and landscape greening. All wild species of *Cycas* are classified under State Protection Category I in China. Illegal excavation and trade and habitat loss are the major endangered factors of existing cycads in China, especially for those with high ornamental value, such as *Cycas debaoensis*, *C. multipinnata*, *C. bifida*, and *C. hongheensis*. Fortunately, the Chinese government issued a series of policies and regulations and carried out a number of key programs for wild cycad species protection and reintroduction, which played a significant role in conservation of *Cycas* species in China. To date, there are only four *Cycas* species (*C. debaoensis*, *C. fairylakea*, *C. panzhihuaensis*, *C. diannanensis*) that have been reintroduced to the wild in China. Among these *Cycas* species, the reintroduction of *C. debaoensis* is rather successful.

2 Description of Reintroduction of Wild *Cycas* Species in China

2.1 Cycas debaoensis*: Feasibility*

There were about 2000 wild *C. debaoensis* individuals, scattered and distributed among 16 populations in 1997 (Fig. 1). The distribution sites of *C. debaoensis* are mostly belonged to the village-owned lands and are in the economically backward regions in which local people have a weak awareness of ecological protection. Due to the illegal poaching of *C. debaoensis* plants and the destruction of their habitats, there existed only 600 wild plants in 2006. Some populations of *C. debaoensis* even have only a few or a dozen plants, which caused difficulties to carry out in situ conservation. In 2007, the State Forestry Administration of China funded and launched the reintroduction of *C. debaoensis* to nature project at the National Cycad Germplasm Resources Conservation Center in Shenzhen Fairylake Botanical Garden, the Chinese Academy of Sciences. Based on the investigation and research on the existing 16 populations of *C. debaoensis*, the project aims to screen 500 seedlings that can better represent the genetic diversity of wild *C. debaoensis* plants based on the resistance gene analog polymorphism (RGAP) analysis. We have also studied the distribution, population size, population ecology, reproductive biology, and conservation genetics of *C. debaoensis* since 1997 (Xie et al. 2005; Gong and Gong 2016; Luo et al. 2014; Wang et al. 2018) and the seed germination tests at Shenzhen Fairylake Botanical Garden and South China Botanical Garden during 1998–2003 and obtained more than 800 young individuals. We established

Fig. 1 The individual of *Cycas debaoensis*

reintroduction sites at Huanglian Mountain Nature Reserve of Guangxi Province in 2007.

2.2 Implementation

In April 2008, the 500 seedlings of *C. debaoensis* were reintroduced to the Huanglian Mountain Nature Reserve of Guangxi Province, where it is less than 10 km away from the type population of *C. debaoensis* (Fuping) and has similar ecological environment (Fig. 2). Thus, reintroduction of *Cycas* species was undertaken for the first time in the Chinese history (Tang et al. 2011).

2.3 Post-planting Monitoring

After transplantation, the survival rate, height, and crown of all transplants were monitored from 2008 to 2018 (Luo et al. 2014; Wang et al. 2018). The follow-up survey after reintroduction showed that, the 500 reintroduced *C. debaoensis* seedlings grew well, of which the leafing rate was higher than 94% after 1 year. The seedling roots strongly developed, accompanied by a certain amount of coral roots, and their survival rate reached 100%. In 2014, the experts agreed the success of *C. debaoensis* in returning to nature, which opened up a new stage in China of

Fig. 2 Reintroduction of *Cycas debaoensis*

successful reintroduction of rare and endangered plants. The implementation of the project aimed to protect and expand the wild population of *C. debaoensis*, to arouse public awareness of the protection of endangered plants and their habitats on which they depend, as well as to explore ways suitable for China's national conditions to protect rare and endangered populations. The implementation results of the project indicated that it is completely feasible to restore cycad population through human reintroduction and suitable tending measures, as long as the problems of habit destruction and human excavation are eliminated.

As a landmark event, the launch of reintroduction of *C. debaoensis* project marks a new phase of rare and endangered wild plant protection in China. The conservation work has developed from simply in situ or ex situ protection to a combination of both, with ex situ protection to promote in situ conservation. The reintroduction of *C. debaoensis* project not only helps to protect rare and endangered species but also directs the public's attention to how to protect the plant species and the ecological environment on which it depends. Therefore, the significance of the present project has gone far beyond the protection of a single species.

2.4 Cycas fairylakea*: Feasibility*

There are two original populations of *C. fairylakea* with more than 2000 individuals in Shenzhen, Guangdong Province. Afterward, the highway construction resulted in severe damage of the Tanglangshan population, of which the wild plants were almost destroyed and some were stolen and sold (Fig. 3). The Meilin Reservoir population was well protected, but it was still influenced by habitat destruction,

Fig. 3 An individual of endangered *Cycas fairylakea*

intensification of pests and diseases, artificial burglary, and low fecundity due to male decline. By 2009, there were about 1000 individuals of *C. fairylakea* in the Meilin Reservoir population, but the situation of their survival was far from optimistic. Most of the plants poorly grow and no plants bloomed. We studied the distribution, population size, population ecology, reproductive biology, and conservation genetics of *C. fairylakea* since 2000 (Jian et al. 2005, 2006a, b).

2.5 Implementation

Since 2010, Meilin Reservoir Reserve cooperated with Shenzhen Fairylake Botanical Garden and conducted a series of rescue and tending measures, e.g., improving lighting conditions by thinning, pruning and cleaning plants in the forest (Fig. 4), strengthening pest control and artificial pollination, and improving nutrition condition by fertilization. This enabled the plants to grow fast and promoted the appearance of more flowering and fruiting plants. At the same time, a seedling breeding center was established, and the harvested seeds were artificially sown to produce seedlings of *C. fairylakea* (most of the seeds were planted in situ in the population).

2.6 Post-planting Monitoring

After recovery of Meilin population of *C. fairylakea*, we monitored the survival rate and blooming individuals of this cycad population from 2008 to 2012. We found the

Fig. 4 The habitat of *Cycas fairylakea* was invaded by alien weed

population had expanded to more than 6000 individuals by 2018, most of which were seedlings that have sprouted in recent years. In the next step, we plan to select robust seedlings to be planted in the nearby Tanglangshan population or other suitable habitats, so as to achieve partial reintroduction.

2.7 Cycas panzhihuaensis*: Feasibility*

The distribution, population size, population ecology, and reproductive biology of *C. panzhihuaensis* were studied since 1996 (Yang et al. 1997).

2.8 Implementation

In 1996, the Nature Reserve Administration of *C. panzhihuaensis* launched artificial pollination, seedling breeding, and reintroduction of *C. panzhihuaensis* plants. We reproduced more than 10,000 young seedlings of *C. panzhihuaensis* (Fig. 5). Up to now, there are about 10,000 *C. panzhihuaensis* plants that have been reintroduced to the wild, which expanded the population to more than 380,000 individuals. Due to severe damage of habitat, there were only 338 wild individuals *C. panzhihuaensis* that existed in Pudu River area. Afterward, the Kunming Institute of Botany,

Fig. 5 The nurtured young seedlings of *Cycas panzhihuaensis*

Fig. 6 Reintroduction of *Cycas panzhihuaensis*

Chinese Academy of Sciences has reintroduced more than 1050 individuals of *C. panzhihuaensis* to the natural population, which enhanced the gradually restoration of this population (Fig. 6).

2.9 *Post-planting Monitoring*

After transplantation, the survival rate and height of transplants in these two populations were monitored. Until now, all transplants grow well (Yang and Chen 2014).

2.10 **Cycas diannanensis***: Feasibility*

Cycas diannanensis is an endemic species at the Red River Basin area in Yunnan. The cultivation of tropical cash crops resulted in severe destruction of the ecological environment of *C. diannanensis*. In 2013, the total number of *C. diannanensis* is less than 10,000. We have studied the distribution, population size, population ecology, and reproductive biology of *C. diannanensis* since 2005 (Liu et al. 2015).

2.11 *Implementation*

Since 2013, the Kunming Institute of Botany, Chinese Academy of Sciences, Daweishan National Nature Reserve Administration in Pingbian, and Dinosaur River Provincial Nature Reserve Administration in Shuangbo County have successively launched related project of artificial pollination, seedling breeding, and reintroduction of *C. diannanensis* (Fig. 7). More than 800,360, and 250 individuals of the *C. diannanensis* have been reintroduced to the Dinosaur River Provincial Nature Reserve, Daweishan National Nature Reserve in Pingbian, and Zhicun Forest Farm in Honghe Prefecture, Yunnan Province, respectively.

2.12 *Post-planting Monitoring*

After transplantation, we monitored the survival rate and height of transplants in these three populations and found all transplants growing well.

3 Problems and Recommendations

It is essential to make an integrated species recovery plan that includes patrolling to prevent plant removal, establishing an ex situ living collection that contains the entire or almost wild genetic diversity, and implementing reintroduction and augmentation to increase population number and size for all *Cycas* plants.

Fig. 7 The nurtured young seedlings of *Cycas diannanensis*

References

Y.Q. Gong, X. Gong, Pollen-mediated gene flow promotes low nuclear genetic differentiation among populations of *Cycas debaoensis* (Cycadaceae). Tree Genet. Genomes **12**, 93 (2016)

S.G. Jian, N. Liu, Z.Z. Gao, et al., Biological characteristics of wild *Cycas fairylakea* population in Guangdong Province, China. Acta Sci. Nat. Univ. Sunyats. **44**(6), 97–100 (2005)

S.G. Jian, N. Liu, Z.Z. Gao, et al., Biological characteristics of wild *Cycas fairylakea* population in Guangdong Province, China. Front. Biol. China **1**(4), 430–433 (2006b)

S.G. Jian, Y. Zhong, N. Liu, et al., Genetic variation in the endangered endemic species *Cycas fairylakea* in China and implications for conservation. Biodivers. Conserv. **15**, 1681–1694 (2006a)

J. Liu, W. Zhou, X. Gong, Species delimitation, genetic diversity and population historical dynamics of *Cycas diannanensis* (Cycadaceae) occurring sympatrically in the Red River region of China. Front. Plant Sci. **6**, 696 (2015)

W.H. Luo, W.X. Tang, S.X. Huang, et al., *Ex situ* conservation of *Cycas debaoensis*: A rare and endangered plant. J. Zhejiang Agric. For. Univ. **31**(5), 812–816 (2014)

W. Tang, N. Liu, A. Lindstrom, et al. *Cycas debaoensis* conservation project in China. Paper presented at the Ninth International Conference on Cycad Biology (2011)

Y.H. Wang, J.J. Gan, T. Chen, et al., Preliminary study on reproductive features of reintroduction population of *Cycas debaoensis*. Subtrop. Plant Sci. **47**(2), 134–139 (2018)

J.G. Xie, S.G. Jian, N. Liu, Genetic variation in the endemic plant *Cycas debaoensis* based on ISSR analysis. Aust. J. Bot. **53**(2), 141–145 (2005)

S.Y. Yang, S.Q. Wan, Z.F. Tang, et al., The artificial forest restoration of *Cycas panzhihuaensis*. China Acad. J. Electron. Publ. House **4**, 18–21 (1997)

Y. Yang, J.R. Chen, The reintroduction of *Cycas panzhihuaensis*. Life World **272**, 44–45 (2014)

The manufacturer's authorised representative in the EU is Springer Nature Customer Service Centre GmbH, Europaplatz 3, 69115 Heidelberg, Germany. If you have any concerns regarding our products, please contact ProductSafety@springernature.com

Printed and bound by CPI Group (UK) Ltd, Croydon, CR0 4YY
30/09/2026
02225333-0007